To Don and Donna.

19 F 95

Francis Whitaker

MY LIFE AS AN ARTIST BLACKSMITH

By
Francis Whitaker
as told to
Gary Mansfield

ISBN 0-9646389-0-8

Lithographed in the United States of America by
LithoCraft, Inc., Boise, ID 83702

F

CONTENTS

THE MAN BEHIND THE HAMMER

I first met Francis Whitaker at his slide presentation at the Museum of Western Colorado in Grand Junction. To say I was impressed with his work would be a gross understatement, however I think it was his personality and his philosophy of life that impressed me most.

That night after I went home, I couldn't get Francis out of my mind. The love and dedication that he had for his work was truly inspiring for all of us no matter what field we are in.

A few days later I wrote a letter to Francis suggesting that he and I collaborate in the writing of his life story. He responded very quickly and positively and soon we were sitting in my office sketching out the details of his early life.

I had recently retired after thirty-four years of public school teaching in the field of writing, and I wanted to get involved in a writing project that would be worthwhile and inspiring to others. I couldn't have picked a better person for this project. At that time I knew hardly anything of the work of Francis Whitaker and even less about the man. In the two years we have worked on this text, I have come to know a man who has become, and will continue to be, a great inspiration to thousands of others as well as to me personally.

If the legacy of the life's work of Francis Whitaker were not enough of a gift to mankind, I think his greater gift would be his patience, his personal philosophy of life, his inspiration, and most of all his encouragement by example to others to strive for perfection in everything they do.

Francis is a great man—had he been in any other profession, he would still be a great man.

There is a little of Francis in all of us—too bad there isn't more!

I hope this book will prove to be as much of an inspiration to others as it has been to me.

—GARY MANSFIELD

Taken from the *New Yorker* magazine, August 8, 1983:

"Our Far-Flung Correspondents"

Frimbo's Peak

by Tony Hiss

(Tribute to Rogers E.M. Whitaker, Correspondent to the New Yorker *magazine, Noted Railroad Buff, and brother to Francis Whitaker.)*

Francis Whitaker, who has been a blacksmith for sixty-one years, invited me out to Colorado a while back to help him fasten a brass plaque he had made in honor of his brother Rogers to a railroad tie ten thousand and fifteen feet above sea level. The plaque, fourteen and a half inches long and six inches high, says:

E.M. Frimbo
Worlds' Greatest Railroad Buff
Rogers E.M. Whitaker
January 15, 1900–May 11, 1981
Traveled 2,748,674.73 Miles
on Railroads . . . July 19, 1981

My first thought when I saw this plaque was that Rogers Whitaker, who for many years was my friend, colleague, and office mate at the *New Yorker* (where he had been nicknamed Frimbo), and whose rail trips I had occasionally chronicled in "The Talk of the Town" had been even more extraordinary than the inscription indicated, because I knew that his Frimbo mileage—which can be divided out to more than fifty thousand miles a year for more than fifty years—was all done on nights, holidays, vacations, and weekends away from his work editing manuscripts, reviewing nightclub acts, and writing a football column for this magazine. It was after I had spent a few days with Francis Whitaker that I began to discover how many different stories, each one as remarkable or as commonplace as Rogers' own story, come together in the plaque that honors him. Since then, back in New York, I have also been thinking about how much has had to survive and remain available during the last half century for even a small plaque with Frimbo's name on it to be exactly where it is now.

Frimbo's plaque is in Cumbres Pass, a narrow gap in the San Juan Mountains, a range of jagged peaks, nine to fourteen thousand feet high, in the southern Rockies. The range is about a hundred and fifty miles long and runs gradually downhill to the southeast into New Mexico. It is empty country and poor country, for the most part, and very beautiful. The mountains are thirty-five-million-year-old volcanoes, scraped by glaciers that disappeared twelve thousand years ago. Cumbres Pass, a couple of miles north of the New Mexico border, is a small meadow with a few tall, straight Englemann spruce trees which is crossed by three paths: State Highway 17, a two-lane road that was paved about twenty years ago; the ghost of an old dirt road used by miners and homesteaders in the 1870s; and a single narrow-gauge railroad track, now over a hundred years old and only about twice the width of a cow path.

There are four seasons in the high mountains, Coloradoans say—winter, spring, fall, and winter. Early in the last week of September, the aspen groves on the slopes below Cumbres turn a pale gold. A couple of weeks later, the trees drop their leaves. During the double winter, from late October until about the last week in May, Cumbres—except for the state highway, which is regularly plowed—is entirely covered with the dry, light, blowing, drifting snow that falls at altitudes of eight thousand feet and higher. There are generally eight or nine feet of snow on Cumbres all winter. One winter, a total of forty-one feet fell; the largest single snowstorm on record left three and a half feet in less than twenty-four hours. In the Cumbres spring, when the winds are gentle, Black Angus and red-and-white Hereford cattle crop lush green grass; towering thunderheads form a white line of mock mountains along the peaks in the clear blue afternoon sky; and, near Frimbo's plaque, purple, blue, bright red, yellow, and white wild flowers—asters, lupine, fireweed, miniature sunflowers, and Indian paintbrush, which is usually fiery red but paled here by the subalpine altitude—bloom among gray-black cinders cast aside by panting little fifty-eight-year-old steam locomotives that clamber up and down the railroad track over the pass.

Driving through Cumbres Pass—the first and less spectacular of two passes on an hour-long trip over an empty road between the small town of Chama, New Mexico, and the equally small town of Antonito, Colorado—is an experience that barely gets into short-term memory. At Cumbres, there's a sensation that a slight rise in the road has just become a slight dip in the road, and there's a road sign that says "Cumbres Pass, Elev. 10022 Feet" (the highway climbs seven feet higher up the pass than the railroad), but there aren't any vistas or panoramas. Cumbres is hemmed in by hills, so you don't think about braking to a stop and rolling down the window, and if it occurred to you to look over your shoulder you'd be gone before you had even turned around. There are a hundred and thirty-six named passes in Colorado, and a car can get over half of them without strain, at least in the summer.

An arrival at Cumbres at a speed of ten miles an hour behind one of the *I-think-I-can, I-think-I-can* steam engines of the Cumbres & Toltec Scenic Railroad is, however, an achievement. It's not something you can do all the time—the trains run only from the first week in June through the second week of October. When they do run, there's one train a day for Cumbres; it leaves Chama at ten in the morning. Standard-gauge railroad track is four feet eight and a half inches wide—an arbitrary but substantial figure that some say is based on the distance between the ruts worn by chariots and wagons in the stones of the old Roman roads of Britain. The rails on the Cumbres & Toltec Scenic track are only three feet apart, and the ground has been cleared for only a few feet on either side of the track. To ride over track of this width is fascinating: it has as much immediacy as a bicycle ride, but you seem to be gliding just above the ground in a dream. Cumbres is an hour away from Chama by train, and it's generally ten degrees colder. The whole trip, twisting through mountain valleys, past stands of pine, cottonwood, and aspen, along ledges cut into the bulging sides of mountains is up a grade so steep that the engine repeatedly throws off roiling columns of gray-black smoke sixty feet high, like a bottle releasing a different genie every second. The engine is often as much as sixteen feet higher than the final car on the train it pulls. When the train gets to Cumbres, the first level place it has encountered in an hour, it has to stop, so that the engine can absorb several thousand gallons of water. And when it gets there it has reached what just happens to be the highest point in the country traversed by passenger trains. So there is it—Cumbres. Frimbo's peak.

Cumbres was certainly never planned to be the highest point railroad passengers could reach, and the experience of Cumbres was not thought by most travelers to have anything to do with their purpose in travelling. Cumbres was just one more thing to put behind you. The sixty-four-mile track between Antonito and Chama was installed in the early 1880s and was only a small segment of a much longer line, now mostly gone, that linked Denver with the towns of Durango and Silverton, way west in the San Juans, where in the 1870s gold and silver had been discovered. There was a small station at Cumbres—there still is. There was even a post office there until 1937, which was really just for railroad workers—the grade up from Chama was so steep that sometimes as many as three extra engines had to be added on in Chama and cut out at Cumbres. But the miner's race to the mining camps was not the only race across Cumbres: the railroad itself, the Denver & Rio Grande, was racing to reach Durango before its rival, the Santa Fe Railroad, could run tracks there. Thus, for a long time, Cumbres wasn't the highest point for passengers, because there were several thousand miles of narrow-gauge track all over the Colorado Rockies, reaching many parts higher than Cumbres (narrow-gauge track was cheaper to build than standard-gauge, and could make sharper turns through the mountains). Then for a time, after all the narrow-gauge track that reached points higher than Cumbres was only *potentially* the highest point for passengers, because only freight trains ran over the line. In 1951, Alfred E. Perlman, the general manager of the Denver & Rio Grande Western, who wished to close down the line to Durango altogether, had discontinued passenger service on it—passenger service that had included, daily, an open-platform car, beloved to Frimbo, where you could get a steak dinner for a dollar and fifteen cents—because without the daily passenger trains, snow would accumulate more often at Cumbres and it would cost more to clear it away, and this increase in cost would make it easier for the railroad to apply to the Interstate Commerce Commission for permission to abandon the whole Durango line as a losing venture. (Perlman, who later became the president of the New York Central Railroad, used similar tricks—except when foiled by Frimbo and some of his friends—to close Eastern branch lines.) But after the passenger trains had been withdrawn, the Durango line started making money again. In the 1950s, there was a huge discovery of oil and gas at the far end of the line, near Farmington, New Mexico, and the pipes needed to bring them out of the ground had to be brought in on railroad flatcars.

Cumbres finally became the highest point for passengers only thirteen years ago, when the track from Antonito to Chama no longer needed to supply the oil and gas fields, was saved by preservationists from Colorado and New Mexico who persuaded the two states to set up the only two-state-owned railroad in the country: the Cumbres & Toltec Scenic, a tourist railroad that would inherit much of the Denver & Rio Grande Western narrow gauge freight equipment, thus preserving the most spectacular part of a working narrow-gauge mountain railroad. No one knows how long Cumbres will go on being the highest point, because the Cumbres & Toltec Scenic is expensive to operate and, although registered as a National Historic Landmark, hard to publicize in the middle of mostly uninhabited country, and a great many people have never heard of it. Kyle Railways, Inc., the private operating company for the line, is hoping to put things on a stable basis by tapping the vast market of Texans who would drive to the railroad's two terminuses on their vacations. (There are eleven Texans for every New Mexican or for every two Coloradoans.) But there it is, for now Cumbres. And you can get there on a train—part of the time, at least. Even if you have to drive to get to the train.

I had Frimbo's plaque on my lap when I heard most of this information about the high pass. It was the morning of the day that the plaque was going to be put in place, and I was sitting on the train to Cumbres, between Francis Whitaker and his wife, Portia. I had already had a day with the Whitakers, meeting them in Santa Fe, then driving to Chama early the next morning—the Whitakers get up at six. Now there were seven of us in the group that was to install the plaque. Besides the Whitakers and myself, there were two other people from New York and a couple from California. We were all sitting in the lead car of a twelve-car train made up of eighty-year-old boxcars with windows chopped out of their wooden sides and comfortable-enough molded-plastic seats bolted to their floors. There were two hundred and fifty people on the train. Francis was wearing a Tyrolian hat and an old Burberry of his brother's—a coat Rogers had had with him for at least a million miles of his train travels. Francis had a picnic hamper on his lap. Portia was wearing a large straw hat and carrying a Polaroid. She told me the names of the wild flowers we were passing. The conductor, talking over a loudspeaker, told an unknown passenger to "put your child's legs back inside the train."

When no one else was talking, Francis, in a quiet, measured voice, told me about the people who had climbed over Cumbres through the years. The mountain sun poured in the open windows and set the coattails of Francis' Burberry flapping. I was very conscious of holding tightly to the plaque. Utes and Apache, Francis began, were probably the first people to find Cumbres. After them came Spanish explorers from Mexico, who named both the mountains and the pass—"Cumbres" is Spanish for "summits," "high-lights," "culminations." After them came mountain men and beaver trappers and then Missouri soldiers hunting Indians, and then miners, and then homesteaders, with their cattle and covered wagons. But more memorable than any of these, to Francis' way of thinking, were the railroad builders, a gang of Irish and French-Canadian workers, who, using only dynamite and hand tools, graded the line, blasted a curving tunnel three hundred and sixty-six feet long out of solid rock, built two wooden trestles more than a hundred feet high, and then skirting a six-hundred-foot drop into Toltec Gorge and cresting Cumbres itself, put down the sixty-four miles of track between Antonito and Chama—and did all this work in just one month more than a year. The first rails on the Chama side of Cumbres, along the grade we were now ascending, were spiked to hand-hewn ties during the snows of November and December, 1880.

. . . "They were rugged men," Francis said. "Is there work still visible? Can we look at the track here now and see how much effort, how much ingenuity, how much firmness of purpose went into it? I often think if we could look at a house or a fence or a garden and see the presence of these things within, then there wouldn't be any vandalism, because we would have too much respect."

While Francis Whitaker was passing on these stories and reflections, I found myself thinking, for the first time of how this expedition had come about. Rogers Whitaker had traveled over every mile of passenger track in North America. But it was Francis Whitaker who had decided on this spot to remember him in. And it was Francis Whitaker who had made the plaque, who had arranged with officials in Colorado and New Mexico to let him place it on their railroad, and who had gathered our group together and put us on the train. I began by thinking, *What a remarkable pair of Brothers!* And it also seemed to me—although I couldn't really explain it—that in getting to know Francis I was getting close to something in Frimbo himself that I hadn't known in all of my years with him.

Francis Whitaker is seventy-six and lives and works in Aspen, Colorado, when he isn't out on the road. He moved to Aspen from Carmel, California, in 1963—even though he was a member of the Carmel City Council at the time—because he thought Carmel was getting too stodgy. He has his own forge, the Mountain Forge, in downtown Aspen, where, either alone or with a student, he hand forges wrought iron gates, ornamental fences, railings, tables, lamps, candlesticks, fireplace tools, and what is called door hardware, meaning latches, hinges, and handles. Wrought iron means worked iron, as distinguished from cast iron, which gets it shape in a mold. People who do this kind of handwork are nowadays often called artist-blacksmiths, to differentiate them from farriers—the blacksmiths who shoe horses—but Francis says he was always happy just to be known as a dirty old blacksmith. For fifty years, Francis has had as much work as he can handle, but he is also now in constant demand as a teacher, and he and Portia regularly travel to schools and forges all over the country, where Francis holds workshops for groups of up to twenty master and apprentice blacksmiths and also gives demonstrations for audiences of several hundred people. Every year, a couple of dozen of his students write him thank-you notes. A recent one said. "I find that your direct blacksmithing approach, supported by your highly developed skills has proven to me that productivity lies in the hands of the smith, not in an arc welder or machine tools, etc."

The skills that Francis acquired as a young man—shaping and putting together pieces of iron by heating them, setting them on an anvil, and then striking them with a hammer while the iron is still hot—had been developed by a succession of blacksmiths over the last four thousand years. A smith is a smiter, someone who knows how to hit metal; and iron is the

black metal: blacksmith. Two thousand years before there were blacksmiths, there were redsmiths—people who shaped the soft red metal—copper. The techniques of the blacksmiths, the product of perhaps a hundred and fifty generations of human effort, observation, and intelligence, almost disappeared in this country during the Second World War, with the adoption of two new tools that almost anyone could master in short order: the acetylene torch, which made cutting a piece of iron clearly as easy as cutting a piece of wood with a jigsaw, and the arc welder, which made sticking together two pieces of iron nearly as simple as gluing them together.

Francis Whitaker taught welding at a San Francisco Shipyard during the Second World War. Then he returned to his forge and went back to work with a hammer. In the late 1960s, there was a revival of interest in traditional blacksmithing in this country, and, in March of 1973, twenty-seven smiths and would be smiths attended a meeting in Lumpkin, Georgia, and formed an international organization, the Artist-Blacksmiths' Association of North America (ABANA)—something blacksmiths in this country had never thought of doing before. Three years later, Francis, who had worked by himself since 1945, and never had exhibited his work except at county fairs, was discovered by ABANA, which by then had six hundred members, and he was invited to demonstrate his methods at an ABANA conference in Carbondale, Illinois. In 1978, Francis served a term as president of ABANA, and since then he has been founding president of the Southwest Artist Blacksmith Association and has been made an honorary life-member of the Upper Midwest Blacksmith's Association, the California Blacksmith's Association, the Arizona Artist Blacksmith's Association, the Southern Ohio Forge and Anvil, and the Appalachia chapter of ABANA. ABANA itself now has more than two thousand members, most of them under forty. ABANA, says Francis, "caught on like a combination of wildfire and snowballs."

Francis Whitaker today is the most sought-after teacher of blacksmithing in the country. His 1983 itinerary has so far included workshops in New Mexico, Alabama, Illinois, Pennsylvania, and North Carolina, where he has helped the Campbell Folk School, in Brasstown, set up a forge with eleven hearths and eleven anvils. He was also head demonstrator at the ABANA tenth-anniversary conference in Madison, Georgia. In between trips, he completed some thirty jobs at his forge, including a four-and-a-half-foot by six-and-a-half-foot gate, a railing to go around the forge, and a hundred and six door handles for a house in Breckenridge, Colorado.

In July of 1981, Francis and Portia added a three-day trip to Cumbres to their 1981 itinerary. They drove down to Chama through Cumbres, spent the night in a motel, and then got on a train for Cumbres the next morning. Francis had Rogers' ashes with him. He had decided to scatter them at Cumbres, but he wanted to bring them there on the train so he could add a final 37.92 miles to Frimbo's millions. Last year, Francis cut the plate for Frimbo's plaque, wrote out the inscription for it, and hammered in the words with letter punches at his forge. Then he set about getting permission from the Colorado and New Mexico authorities to add this piece of hardware to their railroad line and gathered the seven of us together to take the plaque to Cumbres on the train. He told us on the train that he'd had a good time making the plaque, because a brass plaque was one thing he had never made before, and that the timing of this train trip had worked out well for him because he had been able to fit it in between holding a workshop near Santa Fe, where he had made a railing for a circular staircase, and finishing an ornamental fence for a new house in Aspen. (Aspen is by now something of a museum of Whitaker ironwork.) He had already arranged to have the workshop coincide with his annual trip to the Santa Fe Opera with Portia and some friends he has known for fifty years.

"My kid brother has a busy schedule" was what Rogers Whitaker usually said whenever he had a piece of news about Francis to pass on. As a result, when I finally met Francis—two years ago, when he was in New York to visit Rogers in the hospital—I already knew about many of the things he had done. I knew that three years after Francis moved to Aspen he became chairman of the Aspen Planning and Zoning Commission and persuaded the city to adopt a new master plan that put strict limits on future growth in central Aspen. And I knew that Francis had forged a set of iron gates for Leon Uris' house in Aspen using a figure of a Haganah freedom fighter as the center of the design; that Francis had been elected and reelected to the Aspen City Council; that Leon Uris had modeled Conor Larkin, the blacksmith hero of "Trinity," after Francis; that Francis had designed two fireplaces for John Denver's house in Aspen; that Francis had been awarded an honorary doctorate of humane letters by the University of Colorado; that Francis was known as the dean of American blacksmiths. After Frimbo's last visit to Francis and Portia in Aspen, he had reported that when Francis works in his forge he wears steel-rimmed spectacles and a flat black cloth cap; that Francis likes to get to work at seven or seven-thirty in the morning; and he likes to take a break at three in the afternoon and sit on a little bench in front of the forge and drink a can of Coors; that a friend of Francis' caught him doing chores around the house and helping Portia in her garden, and accused him of being a "wife pleaser;" that when Francis had been awarded his honorary degree and was seated in academic robes among his fellow honorees, including a former mayor of Denver and the Chief Justice of the Supreme Court of Colorado, he began waving his arms, trying to attract the attention of his son, Dr. Stephen Whitaker, a professor of chemical engineering at California at Davis, who was in the audience taking photographs and hadn't noticed that he had forgotten to remove his lens cap.

I also knew a few personal details about Francis and Portia. I knew that after Francis met Portia, who is his second wife, seventeen years ago, they got married in two months. (Portia was living in Sterling, Colorado, and Francis told her they were going to get married because their phone bills were higher than either one of them could afford.) That Portia's mother was the first woman to graduate from North Carolina Medical College, married a Russian agronomist, and went off to Russia with him. That Portia was born in Turkestan and that her family had to flee

Russia during the Revolution. That she wound up in Colorado, where she married a building contractor, raised four sons, helped to run the contracting business after her husband died, and learned to ski when she was fifty (Francis can ski two miles down Aspen Mountain in seven minutes flat). That when Francis and Portia got married they didn't know if they wanted a honeymoon, because although they were sure they wanted to live together, they weren't sure whether they would make compatible traveling companions. In the end, they went on a trial honeymoon, driving to one state park, two national parks, and four national monuments in seven days; and when they found that they liked this trip they chartered a sailboat with a captain and a cook and a crew of deckhands for a week in the Grenadines, having first made sure on their final night, when they planned to have steak and champagne in the Tobago Keys, there would be a full moon. Francis is good at planning this kind of moment. He carries in his wallet a picture of himself and two friends taken one morning in California in 1963, at sunrise, on the top of Mt. Whitney, the highest peak in the lower forty-eight states. The three friends had climbed all night by moonlight in order to see the sun rise over the edge of the world.

When I first met Francis and Portia, he was busy dealing with Rogers' affairs, and I really had a chance for only a brief look at him. My first impression was of his *Whitakerness*. There was no doubt that he and Rogers were brothers—they had the same big frame and big head and high forehead and prominent nose. Francis is taller and leaner, but the impression was: *So this is what Rogers would have looked like if he had spent more time outdoors, if he had spent more time standing up, if he had used a hammer instead of a pen.* Francis wears glasses, whereas Rogers didn't, and he has a softer voice and a much milder look; but there was something familiar about the strength inside the voice and the attentiveness behind the look. After Rogers' death, Francis went back to Aspen before I had time to get to know him, and as the year went on, my sense of Rogers himself—strengths and weaknesses both—seemed to be beginning to slip away.

That summer, Rogers' books and papers and personal effects were packed up and shipped to the American Heritage Center of the University of Wyoming, at Laramie, and most of his railroad-magazine subscriptions started to expire. It seemed as though I had just about finished saying goodbye. But it turned out that the good part of the goodbye came a little later, when I began hearing unexpectedly from a few people who had decided to make some kind of memorial tribute to him. John L. Haber, a young New York theater director who conceived and directed "Frimbo," a musical that played in Grand Central Terminal in 1980, began preparing a documentary about Rogers' life. Haber has so far videotaped interviews with forty-six of Frimbo's friends, including night club singers, two Amtrak presidents, a couple of peers, and Harvard's football coach. Robert Lewis, who is the publisher of *Railway Age, Railway Track and Structures,* and *International Railway Journal*, and who had been a traveling companion of Frimbo's since the late 1930s, called up to say that he had somehow persuaded New Jersey Transit to put back in service, for one day only, an old open-platform observation car that Frimbo had ridden in on its very first run, more than fifty-five years before, and that the car would be used for an E.M. Frimbo Memorial trip. (Last August, twenty people, including a timetable publisher and the special projects director of Amtrak, showed up for an early-evening run across New Jersey from Newark to Phillipsburg, on the Delaware River.) And Lewis said I might want to know how many things had been happening in the passenger-train operations since Frimbo's last ride—things that showed that the world might be catching up with Frimbo, that Frimbo had been right never to lose interest in train travel. For one thing, Lewis said, Amtrak had scotched an attempt by the Reagan Administration to reduce it from a nationwide system to a short-haul corridor service between Boston, New York, and Washington. And Amtrak had turned itself into a good railroad, with very modern equipment, trains that ran on time, and a new president—W. Graham Claytor, Jr.—who promised to restore fresh food cooked on board to his dining cars and to replace the plastic knives and forks with stainless steel (Claytor has fulfilled these promises, has added Haagen-Dazs ice cream to the menu, and has even restored over-night shoe-shine service to the sleeping cars of the Florida trains. He has also said that when track work allowing trains from Penn Station to go north via the West Side of Manhattan is completed in a few years, he will name the first train on the new track the Rogers Whitaker Special. There was talk, Lewis continued, of starting high-speed rail service for the first time in this country—maybe in California, maybe in Florida. (Since then, there has ben even more talk—of high-speed magnetic-levitation trains between Chicago and Milwaukee.) Overseas, there were the second-generation high speed trains in France, the restoration of the Orient Express between London and Venice, and the inauguration of a touring train from Delhi, made up of twelve eighty-year-old state coaches with chandeliers, carpets, separate dining cars for Continental and subcontinental cuisine, and an observation with an open veranda and solid-brass pillars.

And I had a visit that fall from Dr. Gene M. Gressley, of the University of Wyoming, who is one of the country's great hunters of archival material, having in the last twenty-seven years obtained the books, papers, and files of eight thousand people for the university's American Heritage Center. It was Gressley who had arranged to have Frimbo's papers and effects from both his office and his apartment—they filled a truck—sent there, he wanted to tell me that the material now formed the Rogers E.M. Whitaker Collection and would enrich the Center's holdings in both transportation and journalism. A few months later Gressley called me to say that a catalogue of the Whitaker Collection—running to three hundred and ninety-four pages, and that thousands of Frimbo's timetables, from six continents and many decades, and twenty thousand of his railroad postcards, together with his correspondence, fourteen bottles of wine, four Bobby Short albums, two homburgs, and a full-dress suit are now available for scholarly inspection in the same building that houses the papers of Ada Louise Huxtable, James Beard, Barbara Stanwyck, and Edward

Everett Horton.

And then I heard from Francis Whitaker. Flying out to New Mexico to meet the Whitakers, I had a sudden sense that Frimbo's memory was being dispersed—to New Jersey, to Wyoming, to Colorado—and I was disheartened. It seemed to risk obliterating a man I had always felt to be all of a piece and one of a kind. The thought disappeared when I met the Whitakers, who gave me a Margarita; treated me to a Mexican dinner in a former convent; took me to the Santa Fe Opera; showed me, on the drive to Chama, one of the oldest irrigation ditches in the United States, which is half a mile long and was dug by Indians seven or eight hundred years ago and has been in continuous use ever since, having deposited many feet of topsoil on the otherwise barren land it waters; and introduced me to a rodeo cowboy who had constructed a machine that swung two drainpipes, like pendulums, in parallel arcs a few inches above the ground. The cowboy was lassoing the pipes in order to practice his steer roping—the pipes moving something like the hind legs of an escaping steer. The cowboy said he had broken his wrist and as a consequence had to relearn his throw using different arm muscles. It took a lot of practice, because first he had to unlearn his old throw—he said that Pavlov had discovered that it takes twenty-nine times as long to break a habit as it does to inculcate one. Francis looked at the pipe-swinging machine and suggested that it would move more smoothly if he added a flywheel to his gear drive. "Oh Portia," said Francis when we were all back in the car, "I'd forgotten what a beautiful ride this is."

It was on the train to Cumbres, as I said, that I first began to put the two brothers, Rogers and Francis, side by side in my mind. When I did this, instead of seeing two hardworking strong-minded, unique, and unrepeatable people, I seemed to see the pair of them at once, moving in tandem across a long stretch of time. It was as if I had just doubled my memories of Rogers. When I thought of Rogers at his desk in his shirt sleeves and braces, or at a football game with his enormous binoculars, or sitting on a train with his Burberry on his lap, the question came to me: *And what was Francis doing then?* After I had spent some time with Francis, he no longer looked to me extraordinarily like his brother, but in his company I felt the same quality of mind at work. I seemed to be noticing not the Whitakerness of Francis but the Frimboness of both Rogers and Francis. I seemed to be seeing, that is, not two sets of life work but one worker making two different series of choices. Or maybe it was one endowment distributed between two different people. So the particular actions of Rogers were just Rogers' way of using something inside. It seemed that such a something could appear in different people, and in people with different interests. But how was such an endowment formed?

The train reached Cumbres, but it was not time for installing the plaque. This was only a quick stop to take on water. The train was heading for a lunch stop at a station called Osier, twelve miles down the other side of the pass. There we would meet the daily train from Antonito, which would pull up alongside us on a short stretch of parallel track. A buffet lunch, choice of American (barbequed beef) or Mexican (enchilada and tamale), would be served to both sets of passengers. After lunch, the two trains would switch engines, and we would get on the Chama-bound train and come back up to Cumbres, where there would be a twenty minute stop and time to install the plaque.

On the slight downgrade into Osier, the engine doesn't work so hard and puts out less smoke, and the trip is quieter, and talking is a simpler matter. I asked Francis if he knew what had led him to forge work, and he told me that the quick answer was his father, and the slow answer was both his parents. He said that his father, an architectural critic who was the first editor of the *Journal of the American Institute of Architects*, in Washington, D.C., was always known in the family as the Skipper, and went on, "In 1922, when I was fifteen and at the end of my junior year of high school, at the School of Organic Education, in Fairhope, Alabama, I came home for the summer and told the Skipper, 'I can figure out how many yards of ditch a man can dig in a day if he can dig so many cubic feet an hour, but I know nothing about his experience, or his life.' The Skipper didn't necessarily see that I had a blacksmith in me, but he was wise enough to steer me in the direction of handwork; and because he knew everyone in the field of building, he got me a place that fall as a helper under Samuel Yellin, of Philadelphia, who made great decorative ironworks for architects, and who was probably America's greatest smith. And the first time I got ahold of a piece of hot iron and a hammer—and I have other smiths say the same thing—I knew this was for me. I never went back to school.

"The fascination was this, and remains this; iron has a strength no other material has, and yet it has in it, at the same time, a capacity for being light, graceful, and beautiful. It has this capacity—but no desire. It will do nothing by itself, except sit there and resist you. Even when it's heated, it remains stubborn and has, at best, the consistency of lead—sort of mushy and sticky or spongy. All the desire, and all the knowledge of how to impart this desire to the iron, must come from the smith. He needs quite a bit of strength, but not an abnormal amount. It is largely upper-body strength, and this is the only reason almost all smiths are men. He has about thirty seconds to work after each heating of the iron—time enough for forty or fifty swings of a two-and-a-half pound hammer. Each blow is doubled, because the anvil underneath exerts an equal and opposite force, but each blow of the hammer affects only the area directly hit. So every blow must tell, and to make it do so, a smith must concentrate all his attention, and be completely relaxed, and not think at all when he swings his hammer. When he swings, he will use his whole body and swing from his heels. And if he gets a drop of molten metal between his hammer-handle and his palm while making a forge weld, he must let it burn and dig it out afterward. The time to think and plan is while the metal is heating. When the iron comes out of the fire, the smith becomes deaf and dumb. Striking is executing what you've planned. I lose myself in my work and get a lot done. Because, when you know how to handle it, iron is very flexible: you can twist it, stretch it, bend it, split it, pierce it, or upset it—which means to thicken it, swell it out. Nothing comes naturally with iron. But when a great smith, like Samuel Yellin, works it, the iron looks as if it could have taken that shape by itself.

"Forge work is a very good combination of the use of the eye, the hand, and the mind—and it's creative. And iron itself is an extremely educational material—it can teach you how to deal with resistant things in yourself, in architects and clients in life. I think now that at fifteen I was starting late—that the human nervous system matures at a rate that makes ten to twelve the right time to begin learning eye-hand-mind coordination. The smith of the old European workshops knew this intuitively, which is why apprentices were started at this age—to help them, not to exploit them. But even at fifteen I could see that there was more than a lifetime of learning here, and I still get such a kick out of it! In sixty years, I've mastered practically every technique that has to do with working iron, but my students teach me things I don't know, and some days I'll figure out a new way of doing something and wonder: *Why didn't I think of that ten years ago?*

"But I began to tell you how both my parents led me to forge work. Before I went to Yellin's shop, I had learned from them how to work. My mother taught us how to cook, darn socks, wash dishes—skills she thought every human being ought to know. She made most of our clothes—even after I left home she made shirts for me for years. She was a dedicated suffragette and got her head cracked once in a demonstration in Washington. Both my parents were perfectionists, in a real way, as long as you were working toward something that was attainable. I've never quite got there with my work, but I can still keep trying, and I have some reason for satisfaction. They gave both Rogers and me a tremendous vocabulary. I think of Rogers as an example of a marvelously trained mind. Rogers always resented the Skipper for yanking him out of Princeton at the end of his sophomore year and sending him off to New York with nothing more than three letters of introduction to New York publishers, where he found his work. But my mother and the Skipper really had already given us the same thing—they made learning interesting for us, and they made work a pleasure. And when I found my work I had good teachers; in my contact with them I had enough real, thorough, basic training to be able to go on and develop my skill. Personal contact—that is the real requirement, the only way things get passed on, whether you are an ironworker or a weaver or a storyteller. I did know a doctor once who taught himself to ski from a book, but when I see someone trying to master something without the benefit of personal instruction I think, *Lord, he'll never make it.* And sometimes I think, *Lord pity us when we come to have an absolutely perfect substitute for that kind of experience. Then we'll have nothing and think we have everything, and that'll be the absolute end.*

Francis' training was a seven-year apprenticeship that was compressed into three years. The first year was with Samuel Yellin, who had created a kind of university of blacksmithing unlike anything else in the history of metalwork in this country. Yellin's name is becoming well-known again; Jack Andrews, a Philadelphia smith and writer, has prepared a major retrospective of Yellin's work, and is also writing a monograph on him. The show opened in Philadelphia last

March. It will move to the Berkshire Museum in Pittsfield, Massachusetts on August 5th, and on October 11th it will come to New York, to the Federal Reserve Bank on Liberty Street. Next January, it will travel to Albany, then down to Wheeling, West Virginia, and then back to upstate New York to Glens Falls and Utica. When Yellin, who was Polish, got to the United States, in 1906, he was already a master of both the techniques and the designs evolved by a hundred and fifty generations of blacksmiths. The traditional designs of the craft are based not only on the smiths' discoveries of what iron can be made to look as if it wanted to do, but also in the shapes—usually bars, rods, or sheets—in which the metal came to the smiths from the ironmakers. The finished product often remains essentially bars, rods, and sheets, but if the smith knows how to bring out what is hidden inside these shapes, then his work looks alive. There is quite a lot of Yellin's work in New York City which anyone can go and see, and after I got back to the city from my trip to Colorado I went and looked at some of it. This was a very interesting thing to do at a moment when its creator was just being rediscovered. All the work I saw looked very much alive, and there were two pieces that even seem to be growing. A pair of massive tripartite lanterns flanking the front entrance to the Federal Reserve

My first major task at Samuel Yellin's. There were 400 of these rosettes on the Federal Reserve Bank job. I made all of them, and think I threw the first twenty away and made more.

Bank and decorated with leaves, spikes, acorns, and small dragon's heads with tongues or teeth, look like elegant, giant, triple-flowered plants that have just burst through the heavy stone blocks of the front wall. At the bank, a vice-president showed me around and took me to the original banking hall, a two-story-high room on the executive floor, so that I could see the fireplace screen and hood that Yellin had forged for the huge stone fireplace, where a fire is lighted once a year for the bank's Christmas dinner. But while the Yellin work in that building is cared for with the respect we accord to a masterpiece, in a second building, work just as good, or better, is treated like second hand furniture. At the old Central Savings Bank Building, at Broadway and Seventy-Third Street, that has recently become a branch of the Apple Bank for Savings, Yellin's work has had to take its chances. Bank employees say they love the ornamental ironwork front of the hundred-foot-long teller's cage, but much of it has been removed to give the banking lines a better look at the tellers; and (at least the day I was there) the delicate little iron roses and flower baskets on top of the cage were dusty and hung with cobwebs.

When Francis Whitaker went to work for Yellin, his shop had sixty forges and two hundred men. Yellin had been awarded the Federal Reserve Bank contract two years earlier; it called for a total of four hundred thousand pounds of wrought iron—the largest single order for ornamental ironwork ever placed with an American blacksmith. To fill his orders, Yellin had taken to meeting immigrant boats and hiring master smiths as they disembarked. He had more smiths with helpers, and the common language was work—eight hours a day, without coffee breaks. In a single year, which Francis says was the equivalent of three years' work nowadays, he worked with smiths from Italy, Germany, England, France, Spain, Poland, Switzerland, and the Virgin Islands. Many of them stayed with Yellin until his death, in 1940. His big commissions had disappeared when construction stopped in the Depression, but he kept his men on salary, paying them with his own money, hoping to set up a permanent school of blacksmithing. He was never able to get this school started. Yellin had a serious heart attack in 1930; he was fifty-five years old when he died.

In 1923, Francis went to Berlin and served a two-year apprenticeship with a German smith, Julius Schramm, who, like Yellin, was famous for his large ornamental pieces. Almost all of Schramm's work was destroyed by bombing during the Second World War. Francis worked nine hours a day, six days a week at Schramm's forge. He says that two years with Schramm was the equivalent of four years' work nowadays. Francis went from Berlin to Carmel, California, where he worked for seven years for a contractor who needed a lot of ironwork because he was building in the Spanish Revival style, which had become popular in the rebuilding of Santa Barbara after an earthquake there in 1925.

In 1932, in the depths of the Depression, Francis decided to go into business for himself. For six months, he welded fenders, sharpened plows, and did whatever work came his way. But by the end of that time he had begun getting enough decorative ironwork commissions to see him through, and it's been that way ever since. It has always been his aim to be his own boss, so that he can meet the people who will be using what he creates; and he has always made it a point to live in a place where work, day-to-day living, and vacations are all a pleasure. He signs everything he makes with a diamond with a capital "F" in the center. This is an ironworker's design which when you open up the diamond, it becomes a "W" indicating the work of Francis Whitaker.

Back on the train after lunch, I was again holding Frimbo's plaque on my lap, and we were slowing down—we were almost at Cumbres again. Then we stopped, and in the silence we heard cows and birds. We saw a cowboy on horseback rounding up seventeen head of Black Angus and also saw two finches chasing each other in front of the train. It was time to install the plaque. Our party of seven got off the train, and Francis found a brand new tie about twenty-five yards in front of the engine, near the place where Frimbo's ashes had been scattered. A brand-new tie won't have to be replaced for some thirty years. Francis rolled up one leg of his pants above the knee, knelt down and took a small drill, eight screws, and a screwdriver out of his pocket, and drilled eight holes in the tie; and then we all gave the screwdriver a few turns. This was just for now; the railroad people had promised Francis that they would bolt the plaque down in a few days. Then we got back on the train, and Francis opened the picnic hamper he had had on his lap during the morning. Inside were plastic champagne glasses, ice cubes, a cocktail shaker, heavy cream, brandy, and creme de cacao. He and Portia had decided the day before that the only fitting way to close the ceremonies was to drink a toast to Rogers in his favorite cocktail—a Brandy Alexander. So that's how the trip ended—with Brandy Alexanders.

It has been said that in midst of all our confusion today Americans are trying to choose the best elements from our culture and set aside less useful things. If we still have a wide range of things to choose from, this is in part thanks to various Frimbos. In the case of the Whitakers, Rogers continued to ride trains and Francis continued to work iron long after most Americans considered both activities irrelevant. When another generation arrived, and wanted to know about such things, the information was still available. And so was the idea that learning can be interesting and work itself a pleasure.

—TONY HISS

REMINISCENCES OF FRANCIS WHITAKER

I was born in Woburn, Massachusetts in 1906. I have very few early memories. My earliest memory was in the flower garden with my mother. We were surrounded by the flower "Sweet William," which I still like very much.

Then I have other memories of being on a farm in Massachusetts with my family. I remember my father being a very versatile man. He loved the land and wanted to farm. He was always an ardent conservationist and his children followed in his footsteps in this.

I can remember running across the road in front of the *one* automobile that might come by *once* a day.

Then memories come back as I recall Boundbrook, New Jersey where we then moved. I can remember some of the events such as the fireworks on the Fourth of July. I can remember the next door neighbors with the noisy kids that we called the "Yaps."

In those days, in Boundbrook, when you went to the grocery store you took your own containers for the goods that you bought.

I remember my mother sending me to get some sugar. I went in, big and important with my sack under my arm, and I helped the grocer lift the lid of the sugar barrel and weigh out the two pounds of sugar that I had been instructed to bring home.

You didn't have the throw away society that we have today.

My brother (Rogers) at that time was twelve years old. He was already a rabid "train nut." He would carry that interest throughout his entire life. I used to go down to the train station with him and watch the many trains come and go. Boundbrook was the junction of three or four major train lines—The Lehigh Valley, the Baltimore and Ohio, the New Jersey Central, and others. I remember my brother, even at that time, could identify all the Pullman cars and what line they were on. He had a real computer mind and could store all kinds of information.

At an early age, I always liked to make things. Toys, Christmas cards, and presents. We made our own Christmas tree decorations. I learned to cross stitch pot holders. There were always chores to do around the house, or at our summer vacation home in Vermont. As I look back on those early days, I realize that I learned more from my parents than I ever did from school. My mother sang, my father was an accomplished pianist, and all four children played one or more musical instruments. My mother taught us all to cook, wash dishes, darn socks, and sew buttons on. She struggled to get women's suffrage and experienced police brutality in working for the cause.

My father was a perfectionist, so when the time came for me to go out in the world, he steered me in a direction that was creative, artistic, and with good financial promise.

I consider myself fortunate to have found my life's work at age fifteen. My parental influence was much more important than I realized at the time. It helped set the pattern for the rest of my life.

1

MY YEAR WITH SAMUEL YELLIN

At the end of my Junior year at a boarding school in Fairhope, Alabama, I went to my father and told him I wanted to end my formal education. I was fifteen years old and it was 1922. I remember my statement at that time:

"Because of my schooling, I now could figure how many feet of ditch a man could dig in an hour and how many cubic feet he could dig in an eight hour day. What I wanted to find out now was how he could live on the two dollars a day that he would make from his labors!"

My father chuckled and said, "It's a tough world out there!"

I had always been big for my age and had always enjoyed working with my hands. And even though I was young, I could fix things, make things, and I enjoyed drawing and mechanics. My father knew that of all his four children, I was the one who should work with his hands. He wanted to direct me into an art or craft where there would be a good future.

My father by that time had remarried and he and my step-mother were living in New Jersey. He continued as the editor of the *Journal of the American Institute of Architects*, and through his contacts with the *Journal* he had met Samuel Yellin and was familiar with the mammoth project at the Federal Reserve Bank Building. He was also well acquainted with the architects in charge. I am sure he was impressed with Yellin and the scope of the project at hand. (The largest ornamental ironwork of this century.) So my father made arrangements for me to go to work in Yellin's shop as a blacksmith's helper.

It was summer time and hot and humid. My father's big black car skimmed over the road to Philadelphia and the event that would change my life forever. But he could not go fast enough for me! I had been basically on my own for a year and a half now and I was anxious to start my life in the adult world. To take home my first paycheck!

When we got to Philadelphia my father got me settled in a typical Philadelphia boarding house. My room was small and on the second floor in the back, but was only a short walk to Yellin's shop on Arch Street. Then my father took me to the shop where I was to work for the next year.

I had not met Samuel Yellin before this time and had never been inside a blacksmith's shop, at least not one this size! All I knew about blacksmiths at that time was that they were usually brawny men who sweated a lot and they put shoes on horses. It was still the horse age and most blacksmiths were spending their time repairing and making things and shoeing horses.

But once I was in this enormous shop my concept of the blacksmith changed forever. It completely fascinated me and has fascinated me to this day. The ringing anvils, the fire and smoke were like the Greek mythology I had read about in school. Sparks were flying everywhere and all around me there were beautiful pieces of iron work being created. My father and Mr. Yellin took me all over the shop. The drafting room, the tool room, the stock room (where I showed the stock man a trick in measuring and cutting iron), then to the assembly room where all the mysterious pieces became a part of a whole. Lastly, the assembled pieces were polished and finished so that their beauty would last forever.

I was in heaven! I thought, *And you mean they are going to* ***pay*** *me to work here?* I would just as easily have **paid** them for allowing me to work in the wondrous shop.

That night I couldn't sleep for the visions of the singing anvils, the dancing sparks, the forged gargoyles, and those humble artists in grimy clothes producing true works of art. Before I fell asleep I vowed to myself, in my little rented room, to devote my life to the art and the craft of the blacksmith. This was the beginning of my life's work and I have stuck to it to this day. More than one thing was forge welded in that shop that day, my determination to be the best blacksmith I could be was forged in my mind.

The next morning finally came and I was off to work. My first job! My father had provided me with some suitable work clothes and with my lunch tucked under my arm filled with some suitable food for a *working man*, I had begun a great adventure, or so I thought.

Even though my father and Samuel Yellin were friends and I enjoyed a somewhat privileged position among the four other boys my age working in the shop, the work was the same. We came in early in the mornings and cleaned the clinkers out of the forges and got the fires started for the day's work. For the rest of the day we were the helpers. We fetched tools, helped lift the large projects, swept the floor, and served as strikers with the different smiths as the need arose. I liked that part best, as I could actually work with the iron and that's what I was the most interested in. Also by moving around the shop as a helper I was able to see the many phases of the work and really begin to appreciate the talents of the men who were working around me.

Yellin had assembled under one roof the greatest blacksmith talent in the United States at that time. Most of the European countries were represented and you could hear all kinds of languages being spoken around the shop. Yellin met the immigration boats and checked the passenger lists for anyone listing their occupation as a blacksmith. If he found someone who offered promise, he hired them on the spot. This is how he could produce such high quality work. He had men who were second and third generation blacksmiths from the old country with the skills and the work ethics of the European craftsmen. This was a combination that was impossible to beat in the United States at that time.

When I started in Yellin's shop, the Federal Reserve Bank project was well under way. The shop was a bee hive of activity and the noise and smoke were sometimes overpowering (this was long before the days of OSHA and other organizations that later controlled working conditions and safety regulations), but I was in heaven and I dove into my responsibilities and duties with a will to learn all I could, and there was plenty to learn!

At the end of the first week I lined up for my first pay check, except in those days you didn't get a check, you got a brown envelope. In my case it had seven dollars in it. I couldn't have been happier if it had been a thousand. I was hooked! With my first pay I paid for my room and board and put the rest in safe keeping for my expenses for the week. I felt very proud of myself and very grown up. I couldn't wait for the weekend to be over so I could get back to work.

I have often felt sorry for people in other professions or jobs who hated their work and couldn't wait for the weekend to come. I have never felt that way my entire life. Probably the most important decision a person can make is the choice of their profession. The work hasn't always been easy for me, but it was always a challenge and I guess that was the key.

Working in the shop with me at times were four other boys my age. They were helpers like I was, but they didn't have the interest in the work that I did. Howard Kaiser was one of the boys, and he and I became very good friends. His friendship helped to ease some of the loneliness that bothered me at first, but soon I made friends with some of the men who worked in the shop and that helped too.

My genuine interest in the work going on also helped to get me some recognition with the older men. Most of them enjoyed their work and realized that they were working on the largest ironwork project in this century (indeed in any century) and the sheer magnitude and the importance of the project was on the minds of everyone involved. The men took pride in their craftsmanship and this made up for the long hours that they put in and the rather harsh working conditions.

With two hundred men working in the shop, and the size and the complexity of the project, the work was necessarily quite specialized. Some smiths worked at the steam hammers, others at smaller power hammers and others did small hand forgings. Each man was judged and assigned

Andirons, circa 1935.

by Yellin on his ability and test pieces were often used to determine this ability.

As the man's ability increased, he was given greater responsibilities and more difficult work to do. Or he might be given the responsibility of supervising the completion of a piece. No matter how many inspections a piece had, Samuel Yellin always made the final inspection before a piece was sent out. He was everywhere. His men had great respect for him because they knew that he not only knew how to draw the piece, but he was a master smith himself.

Yellin made all his draftsmen and architects take a basic blacksmithing course in the shop so that they knew what could be done and what couldn't! This helped considerably in the final product and often I would see the architects on the floor of the shop discussing this or that piece with the blacksmith that would be working on it and the end result would be a collaboration of their putting their heads together and getting the best *and* the most practical design for strength and beauty. Modern industry could take some lessons from Samuel Yellin.

Polished steel screen and firetools to match the marble trim. Private residence, circa 1938.

I feel that the most important lesson that I learned from working at Yellin's was that ironwork, when done right, is a wedding of strength, beauty, function, and longevity. The pieces that were forged in Yellin's shop are just as beautiful and functional today as they were when they were made. That philosophy has remained with me to this day.

The second floor of the big shop was where the final assembly of the larger pieces took place. There were two Italians who worked up there and did nothing but forge animal and gargoyle heads. There were many strange faces up there that looked like they didn't belong to this world, but they all had a place in the grand design.

When all the parts of a piece had been completed the final assembly began. The riveting and collaring of the pieces held them together and the final design emerged. Some final filing and fitting was necessary, but generally the skill of the workmen was so great that a great deal of fixing was not necessary.

I really enjoyed being in the second floor assembly area because you could begin to see the pieces take shape and many of them were truly works of art. When each major piece was completed, white butcher paper was taped to the wall for a background and the piece was photographed by one of the architects who had an interest in photography—and a camera. Usually some of the "boys" as Yellin affectionately called them, would pose with the piece. After the photographs were taken all the men who had worked on the piece would be invited to a small party to celebrate the completion of this phase. There was a great feeling of camaraderie in Yellin's shop. The men really appreciated him and his seemingly divine genius. Yellin included a human form in the photographs to give a perspective as to the size of the piece. He did this on his larger pieces as well, such as the huge gates and entrance doors.

Samuel Yellin kept very careful records, photographs, and drawings of his works and most of these records exist today in the Yellin Museum in Philadelphia.

I was very meticulous and conscientious in my work in those days as I am today, but mistakes do occasionally occur and I was no exception. I guess mistakes are to remind us that we are human after all. Mine concerned a gate frame jig that I was laying out to check the measurements of a gate frame. I made the mistake of measuring from the wrong point and I ended up 1/8 inch off. The man I was working with caught my error and reprimanded me mildly. I was only glad Samuel Yellin hadn't caught me in my first mistake. But as I have always said, "People who don't make mistakes don't get much done!"

The shop at 5520 Arch Street was a fairly large place. It had been designed in 1915 by the firm of Mellow, Meigs and Howe as a blacksmith's shop. Yellin's successes from 1911–1915 had established his reputation beyond all doubt and the commissions were rolling in.

Samuel Yellin's first big commission was to build the gates for the huge estate of J.P. Morgan on Long Island. This was in 1911 and at this point Yellin realized that he must build a much larger organization to handle the larger commissions that were beginning to appear on the horizon.

So from the small three man shop on

409 North Fifth Street, that he had opened in 1909, with his life savings of $150, Yellin now expanded on the strength of J.P. Morgan's commission and other commissions that were coming in.

Yellin was searching the countries of England, Europe, Mexico, even the Virgin Islands for trained smiths who could be recruited into his shop. And in those days if a man didn't or couldn't produce the kind of quality that Yellin wanted, he was let go sometimes that same day. But before that happened Yellin would often work with the man himself, because he was a teacher as well as an artist (he taught nights at the Philadelphia Museum School of Industrial Arts) and if this didn't work out, the man was let go. They didn't have tenure in those days.

By 1915, the shop that I was now working in had been completed and the variety of work was now greatly increased. Weather vanes, business signs, all kinds of door hardware with locks and latches, stair railings, gates of all kinds, candelabras, andirons, fire baskets and huge exterior lamps and brackets, window grilles, all poured from the shop and I was able to see the work daily of the greatest smiths in the world at that time. It was a great experience and I never regretted a moment of the hard work.

This was an international shop. Workers there included: Konetsky, a Polish superintendent in charge of smithing; Boccanera, an Italian master smith, who worked in the shop for forty-five years; Geitz, a German master lock smith; Wincze, an Austrian, in charge of all repousse work; Russo, an Italian, in charge of the polishing department; and Machitti, Fattzzio, and Dezzie, Italian master smiths. Yellin was quite able to communicate with these men, as he spoke three languages fluently, and could communicate about blacksmithing in any language.

The shop of Samuel Yellin was not large by today's factory standards, but for that day and age it was the largest blacksmith's shop in America. Don't forget there were thousands of blacksmith shops just in Pennsylvania at that time, but most of them were one or two man shops and were not doing the kind of work going on at the Yellin shop.

In 1920, there were sixty forges and two hundred men employed in the Arch Street shop. This was a major building period for many cities in the east and Yellin was bidding on most of the big jobs. There were large municipal buildings, private estates for the new and the old millionaires, and private concerns building their home offices. This was the "Golden Age" of Gothic architecture in America and much of the Gothic touch came from the shops of Samuel Yellin. Europe had had her Renaissance and this was really the birth of Gothic architecture in America. Don't forget this was the "Roaring Twenties" and the sky was the limit for many of the companies and individuals that were not being strangled by income taxes and corporate taxes that we have today.

Individuals like Jay Gould, J.P. Morgan, John D. Rockefeller, Henry and Edsel Ford, all were building huge estates that had to have wrought iron gates, window grilles, and other wrought iron trim. It was the perfect time for an iron worker-architect-artist-craftsman like Yellin to reap the harvest, and he did. Samuel Yellin received $260,000 for the Federal

Reserve Bank of New York job which was completed in 1924. In modern dollars that would be the equivalent of over 1.5 million.

As I remember the shop there was a large power shear for making the initial cuts on the larger pieces. Then there were power and steam hammers for the rough shaping of the stock with a power punch for cutting out holes and other shapes, much like a large cookie cutter. There was electric power for some of the motors like the forge blowers, electric drill presses, and some of the grinders. Other power came from a maze of belt drives that were connected to a large shaft that ran the length of the shop. Through a series of reduction pulleys and various levers, the torque of the shaft could be adapted to the work at hand. It was not a particularly safe set-up as the belts some times slipped off or broke, but it was on a good safety level for the day.

Sixty forges lined the outside walls and these had to be maintained by helpers like myself. The power and steam hammers were in the center and handy to the forges. This element of haste in the working of metal hasn't changed a bit to this day. The old adage of "strike while the iron is hot" is just as true today as it was then. Otherwise you are wasting your time!

Designs for both the large and small projects would come from a variety of sources. Some times the architects for the project would be good enough and aware of the beauty and the practicality of good ironwork, so that their designs could be used as they came off the paper. This was not usually the case, however. Yellin might use their basic idea and turn it into a more interesting design which incorporated the skills that he knew he and his men had. His designs were not all original. Yellin maintained a rather large library of work that had been completed in the past two hundred years in Europe and other parts of the world and he drew on this for some of his designs. Copywriting of a piece or a design was unheard of at that time, but there was the natural urge of every artist to be original and Yellin had that urge.

After a design had been formulated, the next step was to make a full size working drawing of the piece. Once the drawing was finished to scale the various workmen would be called in and assignments would be made and the work would begin. Yellin insisted that the full size drawings be posted in full view of all who were working on the project. Yellin's now famous sketches in iron were done to show how his thoughts looked in the real. Each drawing was done with such detail that the smiths could see where to forge weld, placement of the collars, the use of the rivet and the tenon. All these details would be unknown to architects or draftsmen unfamiliar with the skills of the blacksmith. Yellin's drawings allowed men to work together who possibly couldn't even speak the same language.

Samuel Yellin wrote in his article "Iron in Art" in the *Encyclopedia Britannica* 14th Ed. (1940): "Workers in iron should always attempt to make everything direct from a drawing, rather than from models. When working from a model, the object becomes more or less a reproduction, whereas the drawings allow a greater opportunity to express the craftsman's individuality."

As the bigger pieces began to come together, it became necessary to move them sometimes to another location for the final assembly. Although there was a traveling bridge crane that reached across the shop, I can remember many times all available hands being called to move one of the larger pieces. Hand trucks, hands, block and tackle, and the bridge crane were necessary to move some of the larger pieces.

Once the pieces were completely assembled and ready to be installed, they were loaded on trucks or sometimes on rail cars and taken to the site. (All the final inspections by Yellin himself had already taken place.) The final installation of each piece was done by men from the shop and was never left to an outside crew. Anything that Samuel Yellin had his name on was done as perfectly as humanly possible from the initial drawings to the final installation. That was the way he did business and he was very successful at it. I don't recall ever going out on any of the final installations myself.

All the time I was at Yellin's I was like a sponge when it came to absorbing information and techniques. It was a great advantage to be able to move about the shop so freely, as I could watch and help several craftsmen a day. I was eager to learn and the older men were quick to pick up on my eagerness and were very glad to show me anything that I wanted to know—which was everything.

By the end of six months I had proven myself to the superintendent and the various foremen and I was allowed to try my hand at some of the smaller tasks myself. My first and most memorable projects were the rosettes which were on the teller's windows on the Federal Reserve Bank. There were over four hundred of these to be done, as they were at the top and the bottom of each teller's window. And the trick was, of course, that each one had to be exactly like all the others. I remember working very hard on these because they would be the proof to my fellow workers that I did indeed have some talent myself. I labored day after day carefully checking each piece to make sure it was like the others as much as humanly possible. Even today as I am teaching my students and evaluating their work, I will say, "Yes, that is very good, but can you make another one just like it?" I believe that a true craftsman can do this. It is not easy, but if you are really worthy of the name blacksmith, you can.

When they were all finished I was really proud of myself. My work would stand along side that of the other craftsmen in the shop, and be a part of a Samuel Yellin project for many years to come. (It is still there.)

As my year with Samuel Yellin drew to a close I began to realize that if I was going to become a professional smith myself, I needed to expand my experiences and gain more confidence in myself and my abilities.

My year in Yellin's shop had allowed me to see the greatest smith in the world at work. I had watched pieces put together that would stand as milestones for as long as ironwork will be admired in the world. But I was getting restless. I was seventeen years old and I could feel the horizons beckoning me. I wanted to see what other adventures awaited.

The highlight of the year at Yellin's was learning to make a good forge weld. This doesn't sound like a lot, but as good smiths know, the forge weld is the basis of the blacksmithing craft. I remember when I went to Germany the smiths there were surprised that I could forge weld with only one year's training, but that we'll save for the next chapter.

During the year, I had been in contact with my father on many occasions and by the end of the year, he was aware that I was feeling frustrated and confined in my present position. After all, I was just a young helper in the shop and I was showing some promise, but if I was going to realize my potential I was going to have to move to a different environment where I was not just a cog in a wheel.

So at the end of the year, I said goodbye to the friends that I had made in the shop, to my good friend, Howard Kaiser, and to my landlords, the Kellys, at my boarding house and with all my worldly possessions under my arm, I went back to New Jersey to see my father and map out my next move.

Unbeknownst to me, my father had been making some contacts in my behalf. His position with the *Journal of the American Institute of Architects* kept him in contact with architects not only in the United States, but in Europe as well. He had told Dr. Curt Behrendt of me and that I had been working at Yellin's for the past year. I am sure the magic name of Yellin caught the attention of Dr. Behrendt, and soon I was invited to go to Berlin and work as a volunteer (unpaid) at the forge of Julius Schramm.

At first I was a little disturbed at the status of volunteer, because I was always very independent and wanted to pay my own way. But my father explained to me what a wonderful opportunity it would be for me to see Germany and be able to work with the European craftsmen for a while. I reminded him that I had been working with European blacksmiths for the past year, as indeed I had, but he wanted me to go so I did. I never regretted my decision.

(Note: I would like to thank Jack Andrews for the use of his excellent article, "Samuel Yellin, Metalworker" in the Anvil's Ring Magazine*, Summer 1982, Vol. 10, No. 2, which I used as a reference for some of the material in this chapter.)*

2

MY TWO YEARS IN GERMANY

My father had traveled extensively in Europe. He was well acquainted with many European architects through his position at the *Journal*. He once told me that he had crossed the Atlantic by steamer over fifty times. Dr. Curt Behrendt, a Berlin architect, was a friend and it was to him my father first mentioned me and my desire to further my experiences in the blacksmithing field. Through Dr. Behrendt, my father arranged for me to go to Berlin as a "voluntär" or volunteer worker at the forge of Julius Schramm, one of the finest ornamental smiths in Europe at that time.

Finally the day came for me to leave for Germany and I was as excited as my first day at Yellin's. I had never been this far away from home before and moving to a different country for *two years* was certainly going to be a grand adventure.

My father took me to New York to board the *Thuringia,* a small one class boat, for my trip across thc Atlantic to Bremerhaven. The boat was a steamer and *slow!* At least it seemed slow to me, as I was anxious to get to Berlin and get started. It took me twelve days to reach the coast. And to think you can get there in hours today! But the slow trip had one advantage, my father had given me a German dictionary when I left New York and I went over the words carefully every day, while I was aboard ship. I wanted to know at least a little German by the time I reached there. The crew on the boat all spoke German and little English.

At Bremerhaven, I caught the train to Berlin. I didn't realize at the time that Berlin was completely across Germany from where I had landed. The train trip was a real education in itself. The tracks went along the Elbe River Valley and the countryside was beautiful. Old castles and quaint farms dotted the green countryside.

Dr. Behrendt met me at the train in Berlin and took me to his home in Charottenburg. I stayed there for a couple of months while I was becoming familiar with Berlin and my new surroundings.

Dr. Behrendt's wife, Lydia, was a concert pianist and their lavish home was filled with music almost constantly. I had been reared in a home where music was always present also, so this made me feel very much at home. The Behrendts were very kind to me and took me on tours of the city. Dr. Behrendt was an architect and was very familiar with the buildings of this very old and very beautiful city. He knew I would be especially interested in the ironwork—which was some of the best in Europe—so he took special care to show me many of the finest pieces.

Germany was under the Weimar Republic at that time, and it was a lovely place to experience. Although it was a large city and the capital of Germany, it still had the flavor of a small community. There were beautiful parks everywhere and several lakes right in the city, and the people enjoyed life and the outdoors. I often thought of Berlin during the bombings in the Second World War, and how we were destroying a really beautiful city and some really wonderful people.

After the Behrendts had gotten me somewhat familiar with the city, I bought myself a bicycle and joined the thousands of other cyclists in seeing the city and the surrounding countryside. Before long, I met some young American piano students, and we spent many pleasant Sundays exploring Berlin, taking short train trips, swimming in the lakes, and going to concerts. Berlin was a veritable garden of culture with its museums, art galleries, concerts, plays, the opera, and the symphony. I really gained an appreciation of the arts while I was in Berlin, and I have kept this interest throughout my life.

Once, while I was attending a play, I was approached by the management and asked to be a claque during the performance. A claque is a person who is allowed to attend the performance for free if he will clap at specified times to get the audience clapping also. I couldn't have been happier as now I could see many plays for free!

Getting around Berlin was no problem for me, as my German was increasing daily and I had quite a bit of experience in being on my own. The U-Ban (subway) ran from Behrendt's to Schramm's shop so getting to work was no problem.

Shortly after my arrival in Berlin, Dr. Behrendt took me to meet Julius Schramm, the man who would be my mentor for the next two years. The shop was located at Bergmann Strasse 105, in the southwest part of Berlin. His shop was much smaller in comparison to Yellin's. The building was perhaps fifty by one hundred feet. There were only two forges (instead of sixty!) and it was just what I was looking for.

At Schramm's the work force consisted of two master smiths, one master locksmith, and a couple of apprentices. Joining me as a "voluntär" was another young man by the name of Heinz Walz. Neither of us received any pay, but we were covered by health insurance. Heinz and I became good friends and he took me to his home on occasion. He was also interested in the arts and we attended many concerts and went to many museums together.

During my second year in Germany, Heinz and I went on a short trip through Bavaria together. I still have a sketch book of drawings that I made in that delightful country.

But all was not plays, concerts, and sightseeing while I was in Berlin. I was there to work and learn and I did both.

The workday at Schramm's started at 7:00 a.m. We took a break for a second breakfast at 8:30 (unpaid) and then worked until noon. A second break was taken at 3:30 (also unpaid) and then we worked until 5:30. Nine solid hours of work for six days a week.

Everything at Schramm's was hand forged. Unlike Yellin's where there were power hammers for the big jobs. And you can guess who the hands belonged to—me. One of my first assignments was as a striker forging some large newel posts for a stair railing. We were forging 1-1/4 iron down to 3/4 and the job seemed to go on forever. Fortunately, I had quite a bit of this kind of brute work at Yellin's and knew somewhat how to conserve my energy, but this didn't make the sledge any lighter. The forging went on all morning and I could occasionally catch the glances from the other smiths wanting to see how I would hold up under the strain of this continuous pounding. They wanted to see how the soft American would stand up under some real blacksmith work.

By noon I was beat, but the job was only about half over and I knew it was going to be a long afternoon. My shoulders ached and my arms felt like they were going to fall off, but I was determined not to let it show. Heinz and I had a quick lunch and I tried not to think of what was coming in the afternoon.

All too soon the forge blowers were going again and the red hot iron was waiting for me on the anvil. After about an

The German Eagle, designed by Julius Schramm, won first prize in a design competition. I was the blacksmith's helper, making two of them. Body of 1"x6", wingspan 48". They were installed on the Berlin City Hall. Berlin Germany 1924.

hour with my arms and hands throbbing, I realized that if I didn't relax my grip on the sledge hammer handle between heats that I could start pounding quicker when the iron was ready. My hands and arms became a part of that handle for the rest of the afternoon. The problem was, by the end of the job, they had to pry my fingers from around the sledge hammer handle! After that, I was accepted into the shop and the work as an equal and part of the team. The German philosophy was that hard work and sacrifice was good for the soul, the body, and the mind. I had proven myself to them that day—I had won my spurs.

I went home that night tired and sore but with the realization that I had passed the first test. I knew there would be many others, but for the time being, at least, I was being looked upon as having the stuff that good blacksmiths are made of. I felt very proud of myself.

The next day I was so sore that I could hardly raise my arms, but I was back on the job with a renewed vigor. Schramm's shop was very basic. We had electric blowers for the two forges, a drill press, and a tool grinder—that was it. Today I think some blacksmiths starting out think they have to have a whole shop full of tools before they can start. This is simply not true. You can do very excellent work with the simplest of tools, if you have the *desire* to do the job well. This was one of the many lessons that I learned while working at Schramm's and I pass that on to this day. My definition of a good craftsman: One who does fine work with simple tools.

At Schramm's I was working at the forge daily. I learned to forge accurately, to saw and file and fit pieces with precision. This was perhaps the most intensive learning period of my life. I learned to be in constant pursuit of that elusive perfection that was always out there. The German precision and craftsmanship was becoming part of my daily work pattern and I was becoming the kind of craftsman that I wanted to be. My self confidence was increasing and, with the encouragement of my fellow workers and Julius Schramm, I was becoming a *blacksmith* in their eyes as well as mine.

Julius Schramm was a perfectionist. One time, August and I worked hard to make a finial for a large gate. We put the finial in place on the gate and called our master to check our work. "Die bleibt nicht," (that won't stay) he said. He then patiently explained what we had done wrong. This pursuit of perfection has lasted all my life. "One must not only satisfy a client, but one must satisfy oneself." This became my motto.

Schramm's shop was unheated in the winter and drafty and cold. We dressed as warmly as we could and still be able to work. Of course the forges helped to warm the building and to warm us. One advantage a blacksmith has over other workers who must work in the construction or building trades, is that the smith always has the forge to warm his hands, but his feet, those were another matter. Standing on a cold dirt floor in the middle of winter, with no heat in the building, was no laughing matter and sometimes we would all stomp around the shop together to get some of the feeling back into our toes. It was hard, but a learning experience I shall never forget. We all wore wooden shoes, a thick wood sole with a leather part for toes and arch, good insulation.

The workers in Schramm's shop were very enjoyable to work with and we had many hearty laughs together, once they had accepted me.

After two months of living with the Behrendts, I found myself a room near the shop. I missed being a part of their family but I also did enjoy attending the dress rehearsals of the Berlin Philharmonic Orchestra. Wilhelm Furtwangler was the conductor at the time and I had spent many happy Sunday mornings at those rehearsals. At that time, I gained a real appreciation for classical music which I have retained to this day. By now I could speak, write, and read German like a native, and I was becoming quite self-sufficient.

My room included my breakfast and I packed my lunch myself. For supper I ate out as there were lots of inexpensive places to dine within walking distance. The allowance that my father was giving me to live on was quite adequate and with some careful budgeting I could have some of that great German beer with my dinner. I haven't lost my taste for good German beer to this day.

During my first year in Germany, I signed up for some architectural drawing classes at the local high school three nights a week and with this and my daily work I was pretty busy. I wanted to get as much out of my opportunity in Germany as I could. I realized that the time was slipping by rapidly and I would probably not be able to come back at least for several years.

Germany at that time was between World Wars and times were prosperous for most of the people. A number of our commissions at the forge were from the wealthier class. Items such as stair railings, rose arbors, fireplace tools, window grilles, and door hardware were the usual. Also included were some commissions for public buildings. The commission which I remember best was the gates for the Landes Bank. This job was in the shop for six months. I was the blacksmith's helper and I was on the job from start to finish. The gates were masterpieces of the art of the blacksmith. Unfortunately they were destroyed in the bombing of Berlin. I probably have the only photographs of them in existence.

On the smaller jobs I was allowed to initiate much of the design. My architectural drawing at night school was paying off and my designs were considered acceptable by Julius Schramm. This was the kind of experience that I wanted most. To take the initial concept to full scale. To make the full scale drawings, draw them out on metal sheets to check the designs, and finally to make the pieces and fit them together. To me this was the test of the true blacksmith. He should be able to take the idea, sketch it out, draw it to full scale, and then make the pieces and then put them together. The smith that I admired most in the world, Samuel Yellin, had shown me how to do this and now it was up to me to prove my worth.

The final step in any blacksmith's work is the painting of the piece. This may sound trivial, but neither Scramm nor Yellin would allow just anyone to paint their final pieces. Both men insisted that the pieces be painted, at least the prime coat, in the shop. No missed spots, no runs, and no drips were allowed. Anything that had so many hours of design and fabrication in it was not meant to be lost to rust or corrosion. Sadly, many of the works of

Yellin have been seriously damaged by the air pollution in Philadelphia and the same is true of Schramm's works in Berlin. Nothing lasts for ever, but good ironwork comes as close as anything.

During my second year, I was given the opportunity to do some small jobs on my own. One of the last that I did at Schramm's was rewarded with a payment to me. I received fifteen marks ($3.75) for a complete chandelier. It wasn't the money that meant so much to me, but the fact that Julius Schramm thought my work worthy of coming from his shop. The reputation of the mastersmith was at stake, and he was satisfied that I had measured up. Needless to say, I was very proud of myself.

I was increasingly intrigued by the design phase of the whole process. I had seen the excellent drawings of both Yellin and Schramm taken in the hands of masters and made into works of art. However, without the drawing and the design, even the masters would be unable to make true works of art. So I concentrated on improving my drawing and design along with my forging so that I could be the *whole* smith. Even then, I realized that if you were going to build your personal reputation and be able to charge high fees, you had to take the idea and work through the whole process to create the finished product. That's where the money was and that is where the real satisfaction of creation lay.

I tried comparing my designs with Master Schramm's, but his were always superior. However in the process of comparing, I learned many important principles of design. Perhaps the most valuable lesson I remember was when I had designed a large stand to hold a large bowl.

Master Schramm asked me, "How will you support the bowl in the stand?" This was something that I hadn't thought of yet.

Thinking quickly, I answered, "I will put some little supports in where they cannot be seen." I'll never forget his reply.

Sternly he said, "Franz, there is never *anything* that will not be seen!" Words to live by.

As my two year term in Germany came to an end I was filled with mixed emotions. August Klingenberg, the master smith that I had worked with daily at Schramm's, suggested that he and I go into business on our own, but I didn't want to stay in Berlin any longer. I could feel and see the rumblings of the Nazi movement rearing its ugly head. I also could see that it could destroy the country that I had grown to love. I didn't want to stay around and see this happen. Unfortunately, the average citizen of Berlin did not take Hitler and his movement seriously, not seriously enough anyway. They went on with their daily lives unaware of the momentum that was building, a momentum that could and would destroy them.

For my trip home, I booked passage on a little faster boat than the *Thuringia,* that I had come over on. I was a little short of money, so I traveled steerage. My trip was cramped, close, and uncomfortable, but it was fast—five days from coast to coast and I was home. The Statue of Liberty never looked better to an immigrant than it did to me when we pulled into New York harbor.

Immigration was booming at this time and there were hundreds of families on my ship who were leaving their homeland forever. I felt sorry for them, as they looked over the ship's railing at their new home for the first time. They were having some of the feelings that I had when I left Germany. *Am I doing the right thing? What is the right decision to make at this time?* We can all look back at the many crossroads in our lives and wonder what a different turn at that moment would have made in the rest of our lives. But as for myself, I was young, robust, and full of life and I only wanted to know what adventure lay around the *next* corner.

When I reached home, my family hardly recognized me. I had grown another two inches and filled out in the arms and shoulders from all that sledge hammer work at Schramm's. When I first spoke to my mother she said with a shocked look, "Francis, you have a German accent!" I am sure I did. After two years living in Germany, I had picked up the language, the writing, the customs, the work habits, and, of course, the accent!

After a few weeks of getting re-acquainted with my family, I tried to find a shop where I could start applying the skills that I had been learning for the past three years. My father seemed to be quite impressed with what I had learned and with the pictures of the completed works I had worked on in Berlin.

Unfortunately, I was not successful in procuring a shop of my own. My father had gotten some commissions for me from some of his architect friends and I was eager to prove my skills, but we could not locate a shop where I could complete them. I even tried to buy some time in some of the existing shops around New Jersey, but the smiths there were very busy with the horse trade and making things for their own customers. They were not much impressed with the kind of work that I wanted to do anyway, as their work was of a much more practical nature rather than the artistic work that I wanted to do. Decorative ironwork was not much in demand for the common man at that time.

After a month of frustration trying to find a shop for myself, I took a job with a good builder (again a friend of my father). I continued to live at home and began work as a carpenter. This went on for several months and caused my father no end of grief. He felt he had invested three years in starting my career as a *blacksmith* and he was determined that that was what I was going to be. He had seen the magic that Yellin and the European smiths could do and he wanted me to follow in their footsteps.

After several months of my working as a carpenter and living at home, my father announced one evening at the dinner table that I was going to California! I was shocked and a little upset at the time, but I wasn't going to show it. He told me he was going to send me out with one of his architect friends who would help me get settled and started in my own business. He said he didn't want me to waste any more of my training and skills bending nails on a carpentry job! And that was that!

As usual when the "Skipper," a family name for my father, said something it came true with little delay. My father gave me $35 and bought me a *one way* ticket to California. I remember his words when we parted. "My son, from now on, you are on your own." I couldn't wait to see what would be around the next corner in my life.

3

MOVING TO CALIFORNIA 1926–1963

With the *one way* ticket to California in my pocket and the $35 dollars that my father had given me to start my fortune tucked carefully into my wallet inside my coat, I boarded the train to whatever awaited me in the mysterious and far-off west.

My father had been to California several times and knew that he was not sending me to fall off the edge of the earth, but I wasn't so sure. What my father also knew was that the Santa Barbara earthquake had been over for a few short months and the City Council had already decided to rebuild the city in the Spanish tradition. Architects and builders of all sorts were quick to get the news and there was a general migration of building tradesmen to the Santa Barbara area. Of course with my proven blacksmithing and architectural iron working experiencing, I could move in and capitalize on this building boom. If I have anyone to thank for my early success in the trade of blacksmithing, that has to be my father. He had the foresight to place me where I could get the best training in the world at that time, also where I could see the work of the most skilled hands in the trade. At eighteen years of age I had worked with the best and seen the work of the best so now it was up to me to apply what I had learned and carry on the traditions that I had seen in Europe and America.

When I left home, my brother Rogers said, in horror, "You're going out there where all those Indians are?" I tried to reassure him that most of the Indians were fairly civilized by now and that the chances of my getting scalped were fairly small, but he wasn't so sure.

My trip out to California on the train was very pleasant and very educational for me. I had never seen anything that even vaguely resembled the country that I was now in and the deserts, mountains, and beautiful valleys made me feel that I was making the right choice—although the choice was not totally my own. Actually I think my father would have liked to have moved to California at the time too, but his work kept him in the East as that was the hub of the architectural world at that time. Since he couldn't go—he sent me instead. He was very much aware of the potential for growth in California, and as Horace Greeley had said earlier, "Go West, young man," and I was very happy now to do just that.

Riding on the train in the 1920s was a

wonderful experience in itself. Anyone who was anyone rode the trains in those days and the passenger lists were a microcosm of the population of the country. The passengers came from every strata of life at that time and you would see the very rich and those from moderate means—like myself—dining in the same dining car and enjoying the same kind of service. There was some class distinction on the railroads at that time, but not more than in the society of the "Roaring Twenties."

Even today, I travel on the trains wherever possible. It is a wonderful way to see the country and you can see a different side of a town or city by way of the railroads. Some people think that the railroad view of a city is the ugliest side, but I don't agree. It is just the working side of the life of the city. The parts of town that make the other parts work and look good. I guess it's kind of like blacksmithing, you have to get your hands dirty and make some smoke to create something beautiful. I believe the effort and the mess are what make the product worthwhile.

It is easy for me to see why Rogers, my brother, wanted to spend his free hours riding the passenger trains. (Rogers logged 2,748,674.73 miles on passenger trains.) You could glide through the countryside with a seemingly unlimited view of the way people lived. It was kind of like peeking through their windows without their knowing it. And at night the clickety-clack of the rails lulled you to sleep as you watched the glowing lights of the farmhouses—many of them were still lit by oil lamps at that time, and of course, the engines were all steam!

Steam engines added the flavor to train travel that I think was lost with the change to diesel. The wail of the whistle in the night, blowing for some unseen crossing, was music to my ears. I'll never forget the magic of travel by steam locomotive. It was grand and was certainly a page of Americana. AMTRAK is good, but it doesn't compare with the luxury of the good old days of train travel.

When I arrived in Los Angeles, I had no trouble finding a job. The contracts had been let for the rebuilding of Santa Barbara and many of the shops were hard at work. Federal Relief moneys by the millions were pouring in and they were being dispersed almost as rapidly. The stipulation that the funds were to be used to rebuild using only the Spanish style of architecture, was a great advantage to some builders and architects who were familiar with this style. Also it was a great boon to the blacksmiths like myself who would be doing the ironwork for the restorations. Again I had my father to thank for his foresight in sending me to California at this opportune time.

Before settling down in Los Angeles, I decided to visit a high school girl friend that I had known in Alabama. She had moved to Carmel recently and we had been corresponding. When I arrived at the girl's house and told her that I was looking for a job in the area, she told me of a local contractor who had a blacksmith shop and a drunken blacksmith that was giving him a lot of trouble. M.J. Murphy was the name of the contractor and it turned out that he was the largest and best builder in Carmel at that time. The next day I visited Mr. Murphy and after a short interview, I convinced him that I was capable of doing the type of work required for the Spanish restoration that he was doing. He was not thoroughly convinced so I reported to the shop the next day, and after a trial period of three days, I had convinced him that I was the man to take over his shop.

I rented a small room in a boarding house in Carmel, a few blocks from the contractor's shop, and began my first full paying job in three years—in fact the first full paying job I had had in my life. I was feeling pretty smug and pleased with myself by that time. I had ridden a train clear across the country, found myself a girl friend, and gotten myself a job—all in the course of a few days. I knew my father and my family would be proud of me and my progress. But most of all I was proud of myself!

My job with M.J. Murphy lasted for seven years. I did all sorts of work. The shop was owned by Murphy, of course, but I was in charge of all the blacksmith work and the repair work related to the construction business. I sharpened picks, (this was before ditch digging machines and there was a lot of hand labor), I made equipment for the team of horses that were a part of the company too, and although I never shod a horse, there were a lot of other parts and pieces to be made and repaired. I made special tools for forms and scaffolding, as a lot of the work was made of poured concrete finished to look like adobe. For the interiors of the houses, I made curtain rods, gates, railings, door hardware, pot hangers, light fixtures, window grilles, and anything that gave the houses the Spanish look that the people wanted. These were houses for the more-or-less middle income people of Santa Barbara, so most of them were not the mansions of southern California that we sometimes think of.

The years I worked at Murphy's were also the peak of the sardine fishing industry in Monterey. All sorts of tanks were needed for the unloading and storage of the sardines. My blacksmithing skills were called upon to make the "clamp bolts" that were needed to hold together the redwood tanks. I made hundreds of those bolts.

In addition, I was called on to do some other work at the canneries. It was there that I learned a process that was to be invaluable to me for the rest of my life. The process involved bending an angle iron into a circle without distorting it. The canneries were tall buildings, and at the top of the building the empty cans started their journey to the bottom. They were steam cleaned, packed with sardines, steam cooked, the lids were put on, and finally they were labeled and packed in boxes.

The route of the cans went along angle iron tracks which guided them through the various processes. These angle iron tracks had to be made very precisely so the cans wouldn't get caught or fall through the tracks. The angle smith, a specialized branch of the blacksmith profession, was the person responsible for the very complex track system. That was all he did. He built, repaired, and maintained the tracks—probably the most important job in the factory. I learned some tricks about angle iron from that man! The cannery work, and the related blacksmithing were very interesting and unforgettable times in my life.

Unfortunately, the rapid growth of the sardine industry finally resulted in their extinction. Canneries were abandoned, some burned to the water's edge, others finally became restaurants and gift shops for the new industry—tourism.

Things went along very well for a year,

so well in fact that in 1927, I asked my former girl friend to be my wife. When Elaine and I were married, I was twenty years old and she was nineteen. A lot had happened to me in those twenty years and everything indicated a bright future for both of us. The building trade was good, the economy was good, we lived in probably the most beautiful area in California and the future couldn't look brighter. And then it happened.

Two years after Elaine and I were married the Crash of 1929 brought the business climate of the United States to an abrupt halt. Little did anyone, especially those of us living in California, think the disaster on Wall Street, a full continent away, would affect our lives, but it certainly did!

I was working at M.J. Murphy's when I heard of the stock market crash. Since this was the first one that had ever occurred in the United States in this century, we all wondered how it would affect our lives and certainly how soon?

Our construction work for Santa Barbara went on about the same for several months after we received the news of the crash, and at first I thought that maybe New York was too far away to affect us—boy, was I ever wrong!

The bottom didn't drop out completely until 1932. But when it went—it went completely. I was laid off at Murphy's as the end came, and for the first time in my life, I was out of work and had no prospects at all of finding any. It was as if the whole work world had suddenly stopped. There were no jobs of any kind for anyone, and especially in the construction business.

Men walked the streets or stood in long lines for handouts from various government agencies. We read in the papers of the hundreds of suicides of the rich and famous in New York, Chicago, and the many large cities. It is hard today for us to imagine the scope of this disaster. Bread lines, soup lines, and other humiliating experiences were the order of the day for the rich and poor alike. The Roaring Twenties had been the crest of the wave and now came the crash which devastated the economy of the entire country.

I was now twenty-two years old, I had house payments, a family to feed (my first child had just been born) and I had no job and no prospects for one. All the rosiness of our first two years of marriage had suddenly been crashed in the dust. It was a devastation the extent of which I had never known before. People of those times were proud and independent. There had been no welfare system in the United States before the time of the Great Depression. People wanted to stand on their own two feet, but alas, there was nothing to stand on! In Carmel, we had some help from the Red Cross in the form of medical aid and some clothing and some processed foods were available. I took a job with the State Emergency Relief Administration taking a cat and dog census in Carmel, but that didn't last very long. There were not enough cats and dogs to keep me busy, too small a town, I guess.

I took a part-time job in a local dairy washing bottles, but that didn't last very long, as most people didn't have much money to buy milk. I did get free milk for my family though, which helped. From there I worked for a highway construction company. There was not much work, however, but patching the existing roads. A lot of patching was going on during that time all across the country—and it was not all on roads!

Eventually I got on with the U.S. Forest Service as a blacksmith on a road construction project in the Big Sur area. The pay was not so good, but the views were magnificent. I kept saying to myself, "If you have to go broke, there isn't a prettier place in the United States to do it in!" The work was hard and not at all what I wanted, but I had a *job* and that was a lot more than most of the people in the United States could say at that time.

By 1934, things began to lighten up a little. The Building and Loan Company had threatened to take our house, but when they found that they had more houses than they knew what to do with (they sure couldn't sell them) they were satisfied to let us just make the interest payments. This we did for almost two years and by paying them, we saved the equity that we had built up. The times were incredibly tough. I don't think anyone who lived through the Great Depression ever got over it. It was very difficult to believe, after it was over, that you would never have to worry about where the next meal was coming from. It was an experience that I'll never forget, and I certainly wouldn't want to repeat!

John Catlin was a retired lawyer from Sacramento who lived in Carmel and had a small blacksmith shop in his garage. He had set up the shop as a hobby and for something to play around with in his retirement. He had done some smithing on his grandfather's ranch when he was growing up and wanted to take it up again. I had heard of the man through the grape vine—there was a good grape vine system in Carmel in those days because practically everyone knew everyone else and their business as well. My Forest Service job was very seasonal and shaky at best, so I was looking around for a way to get back into ornamental blacksmithing that I had been trained to do.

I went to Catlin's and introduced myself and told him of my background and my training. He was impressed enough to let me come back and help him on some of his larger projects. Even though I was still working full time at Murphy's, I came over on weekends and sometimes after work. Eventually he allowed me to lease his shop and equipment and set myself up in business.

The arrangement was good for both of us. John Catlin was able to benefit from my training and experience and I was able

Andiron, circa 1940.

to use his shop. He called it the Forge in the Forest and that was the name I used when I started in business for myself.

After my first year, Catlin decided to build a larger shop to accommodate the increase in my business. He had some influential friends, architects, and clients and he helped to spread the word of the kind of work I was producing along with the word of mouth of the quality of my work.

It was slow building up a business after the Great Depression. Carmel was not an industrial area and the economy was based largely on the scenery, the fishing in Monterey, and the tourists in the summer. This area of California had not really been discovered by the tourist population as yet and there were not a lot of tourists from outside the state in those years.

I never advertised while I lived in Carmel. The quality of my work spoke for itself and I always had plenty to do. This has always been my theory on the blacksmith advertising his work. First of all, advertising is expensive and you must raise the cost of your product to absorb the cost of the advertising. This makes higher costs and possibly limits your sales. I preferred the mouse trap theory and it has usually worked. Once you can get some of your work displayed or installed in the right homes, more business will soon be forthcoming. If your work, of course, is of the quality that people, who can afford it, want. And it is up to the smith to make sure that his work is of this quality.

Carmel, or Carmel By the Sea, as it is called, was an ideal place to live in the thirties. Over the hill was Monterey, which was first sighted by Don Juan Cabrillo in 1542 when the white beaches and pine forests of Carmel were first seen by white men. Sixty years later Sebastian Vizcaino was the first European to set foot on the peninsula and he dubbed it "Monterey" after a viceroy of New Spain named Conte de Monte Rey. Vizcaino also discovered the valley and named it after his patron saint, Our Lady of Carmel. The Spanish did not colonize the area until fifty years later with the establishment of Alta, California in 1602.

In 1770, Father Serra selected a site near the mouth of the Carmel River to construct the second of what eventually became twenty-one Spanish missions in California. Monterey became the Mexican capital when the Mexicans broke from the yoke of Spanish rule in 1821. In 1846, Commodore John Drake Sloat of the United States Navy claimed the city of Monterey for the United States. In 1846 the California Constitutional Convention was held in Colton Hall (the oldest public building in the state). Later that year California became the thirty-first state of the United States.

Monterey for a time was the whaling capital of the west coast. Whale oil was used to light the lamps of the world before petroleum and electricity.

John Steinbeck's setting for his two great novels, *Cannery Row* and *Sweet Thursday* was born when Cannery Row was built in the 1880s for the processing of sardines. It is a great tourist attraction today.

The 1906 San Francisco earthquake sent many artists, writers, and musicians to Carmel and Monterey to form what was later known as the Nation's Art Capital. The mountains, the sea, and the land were an unsurpassed combination unequaled anywhere in the world and truly an inspiration to anyone who saw them. Point Lobos State Reserve (which I helped to protect) is called the "greatest meeting of land and sea in the world." It was the model for "Spy Glass Hill" in Robert Louis Stevenson's *Treasure Island*. Stevenson was also a resident of Monterey and the Stevenson house is open to the public there.

Living in Carmel during the 1930s was an experience in itself. This small community enacted many restrictions banning neon signs, billboards, and hot dog stands and served as a model for many other small communities that wanted to preserve the small town atmosphere for over two hundred years. The Cliffs, the Big Sur, and

Carmel, California about 1945.

the Giant Redwoods make it one of the most beautiful spots on earth. My family and I were certainly fortunate to live there.

Those early years in Carmel were very pleasant and peaceful. There were a lot of poor artists in the artist's colony, a few rich people, and many summer tourists made up what we called the transient population. All the permanent residents knew each other and the winters were very quiet. In the summer, the people came to the cool coast from the hot interior. Many of them, as the times became better, bought second homes in Monterey and Carmel and spent their vacations and weekends on the coast.

There were many famous artists and writers living among those hills. Some of them hadn't become famous yet and maybe never would, but it was certainly a unique collection of talent.

Jack London, Mary Austin, Albert Rhys Williams, Lincoln Steffens and his wife Ella Winter, John O'Shea, William Ritschel, Fred Bechdolt, all were my neighbors, customers, and sometimes friends. John Steinbeck and Ed Ricketts of Cannery Row fame became close friends of mine. The Carmel Music Society brought famous artists to Carmel for concerts and we had an opportunity to hear and see the best musicians of the times.

Herbert Heron founded the Forest Theater and Ted Kuster founded the famed Golden Bough Theatre both in our area and I became involved in many amateur productions both as an actor and a stage hand. It brought back many memories of my days in Berlin where I guess I really began my interest in the arts and live performances.

The scenery of the area attracted many photographers as well, some of which went on to become famous. Johann Hagemeyer, Edward Weston, and Ansel Adams were my clients and friends.

Much of my early work in Carmel centered around fireplace equipment. Natural gas had not arrived in our area and the mild climate made fireplaces the best choice to take the damp chill out of the house and warm the long winter evenings. I made andirons by the hundreds, some were quite large for the larger homes, and all kinds of fireplace tools. Fireplace screens were also in great demand to keep the sparks off the floor and decorative and practical fireplace hoods sold well.

Our home had, like many others, a hot water coil at the back of the fireplace. I built a fire and waited for hot water for a bath.

My largest commission while I had my shop in Carmel was for the Santa Catalina School in Monterey. They wanted to follow the Spanish motif to the letter because of the historical Spanish influence in the area. I made window grilles for all the windows; door hardware such as handles, latches, hinges and locks for the doors; along with railings and banisters and altar pieces. I prided myself in the great variety that I could come up with and even though there were hundreds of pieces, no two were alike.

The next disaster in my life started on that fateful Sunday morning, December 7, 1941, with the bombing of Pearl Harbor. It was not just my disaster of course, but it did change my life considerably for a time. As soon as the war effort was entirely mobilized, it was impossible for me to get the materials that I needed for my work, so

I was forced to close my shop and find work elsewhere. I had a little experience with Fascism while I lived in Berlin and I knew I wanted to get the war over as quickly as possible. So being too old to join the fighting, I decided to join the effort and build ships with which to fight the war.

I went up to San Francisco to the shipyards and filled out an application. My experience in all phases of welding and metallurgy attracted immediate attention, and I was hired and put in a top paying job right away. The type of work I was doing was completely different from anything that I had ever done before, but it was still working with iron and steel and I had a lot of that. With the daily increasing demands for ships there was lots of overtime and many long hours with little sleep. New men and women—inexperienced—flooded in daily and getting them organized for production was the most difficult task.

The work was a daily challenge to map out the best plan of attack. Having had several years now of mapping out my own plans I was now called upon to work with a much larger force and, of course, on a much larger project. I improvised many ideas to expedite the work and made the best possible use of the equipment and

The Forge in the Forest, Carmel, California. Finishing several large grilles for the Santa Catalina School, Monterey, California. 3/4" square bars, with 1/2"x3/4" diamond pieces forge welded. 3/8"x1" collar forge welded top and bottom with 1/2" tenon. This commission forced me to develop a spring swage to make the tenons. 1960.

manpower that we had available. I enjoyed the work and it certainly was a challenge.

After I was established, my family moved to Mill Valley, across the bay from San Francisco and close to one of the shipyards. San Francisco was a mad house at that time of soldiers, sailors, shipwrights, and a couple of million other people going to or coming from the war.

After a year in the shipyards, I was advanced to teaching in one of the shipyard welding schools. The work in no way was related to blacksmithing as only a few parts of a ship ever were forged, but my general welding experience came in handy and it was a break from the day-to-day work in the yards. The ship yards needed hundreds of welders and there were few available before the war, so the shipyards decided to train their own. Several welding schools were set up to make welders as soon as possible, sometimes a little too soon. Most of the jobs, however, were so specialized that it didn't take the average worker long to learn his job. I wouldn't call these people welders, but they knew how to do enough to fulfill their jobs. We worked incredibly long hours and sometimes weeks at a time without a break. It was a wonder how the men and women held up under the strain.

By 1945, and after four years of building ships, I and my family were ready to go home and continue our lives. I remember vividly where I was when the war was over. I had worked seven weeks straight through in order to get a week off to go backpacking in the Sierra Nevada Mountains with my family and a good friend. We were hiking and camping the entire weekend and were never near a phone or a radio. When we came out of the wilderness, we stopped to get some gas and found that Armistice had been declared. The war was finally over! The filling station attendant couldn't believe that we didn't know it already, but when I explained that we had been in the mountains camping, then he believed us.

We couldn't get out of San Francisco fast enough. My family and I moved back into our house in Carmel and I started building my own "Forge in the Forest." My shop had been closed during the war, but soon the anvil was ringing again.

Thinking back to my trip to California on the train, I remember getting off the train in Kansas City and picking up—from the newsboy—(they had them in those days) an Extra Edition of the newspaper with the headlines "Lindbergh Flies the Atlantic" and "Lindy Lands in Paris!" Sometimes as I think back over my life, it is hard to believe all the changes that have taken place over the years, just within my lifetime—from the last horse drawn street car on Canal Street in New York to the first plane trip across the Atlantic to walking on the moon. It's incredible!

Peacetime soon settled in and it was certainly a pleasant change. With the war over and rationing and shortages of all kinds now at an end, I could get the kind of materials that I needed to get rolling again. The people who came to California for various reasons during the war, stayed and soon there was a population explosion and a building boom. The economy picked up and construction started again. Soon I was busy, but at my *own* forge this time!

Not long after the war, Carmel began to suffer from what I called "growing pains." With the war over and servicemen returning by the thousands and needing new homes and new towns to settle in, the migration out of the big cities began and the era of the suburbs began. This was happening all over the United States, but California received the brunt of the migration. Carmel was one of those incredibly beautiful places that was about to be invaded by the outside world. Workers who had come to California to work in the ship yards and war plants from the South and the Midwest saw what a beautiful state it was to live in, with the wonderful climate the beautiful scenery and plenty of jobs—so they just stayed. Before long they sent for their parents, cousins, brothers and sisters, and on and on it went.

Before long I was becoming concerned about the ramifications of this influx of population on our little community. It was one thing to live among all these people during wartime, but now the war was over and I wanted the peace and quiet of our little hometown back again.

Builders, contractors, and developers were advancing on the Carmel area seemingly by the car loads, and I and many of

Main grille and window guard. Santa Catalina School, Monterey, California. Similar construction as the plain grilles.

the other permanent residents were becoming very worried about the future of our community. There was only one thing to do . . . Get Organized! That is how I became involved in politics. I was probably not the first blacksmith in the history of the United States to get involved in politics, and I probably would not be the last. Blacksmiths by their very nature are: stubborn, independent, fearless, thick-skinned, bull-headed, determined, and they don't like to be crowded, so they make perfect politicians!

I ran for the City Council of Carmel and was elected! And I was re-elected twice. I'm not going to say that my platform was to keep the outsiders out—that would have been impossible—I just wanted some control over the rate and the kind of growth that we were going to have. I had a great deal of support from my friends and the other powers in the community that also wanted some control of our collective futures.

Our first step was to develop a Master Plan for development in and around Carmel. It was a ten year plan which covered everything from storm sewers to regulation of hot dog stands. I was immediately very popular with some people and collectively hated by others. Our first job was to show the haters how a plan of this sort could be beneficial to *all* the residents and the non-residents alike. I guess I was successful as I was re-elected for two more terms.

I also became involved in other conservation types of activities in the area. I was President of the Point Lobos League to help preserve Point Lobos and see that it was made into a State Park. I helped with the acquisition of the Carmel River mouth and the preservation of its beaches for public use. Without my contributions, and that of others, these areas would have been bought up by developers and turned into hotel complexes and other defacers of the beauties of nature, and made into private preserves of the rich. I just couldn't stand by and see this happen. The coast line was such a beautiful place it should be available to be enjoyed by all—not just those who could afford to *buy* it.

It is my opinion even today, that the business people of any community are the people who should be the most concerned with the quality of life in their town or area. As a blacksmith and an independent business man in Carmel, I felt that I should be sensitive to zoning regulations, public parks and beaches, and the myriad of other features that an area has to offer and to make these features available for the general public. After all, the business people are the ones who benefit from a safe, healthy, happy community. Their business grows in this environment.

Although my customers came from all walks of life and levels of the social strata, they all basically wanted the same things; a clean, safe, well-ordered community that was a safe place to raise their children and was a pleasant place to live. I believe it is the mission of the business community and the independent residents to do all within their power to try to maintain this kind of environment. The City Council of Carmel, I believe did this, and I was glad to have been a part of the work that was done. We were a model for many other communities along the coast as well as inland.

My experience on the City Council and the formation of our Master Plan with all its ramifications was excellent experience for me, for after I had moved to Aspen, I was asked to work on the planning commission there also, and I had a pretty good idea what we were getting into. But this is in the next chapter.

My move to Aspen was prompted by several incidents and experiences that were not necessarily related, but the end result was the same. During my years of living and working in Carmel, I experienced a growing interest in the sport of skiing. When I had been a sophomore in high school, our shop teacher told us we must come up with a project to make in class for our grade. Most of the kids were going to make book shelves, candlesticks, magazine racks, and things like that, but I decided that I would make myself a pair of skis. Our teacher thought that this would be too advanced a project for a tenth grade workshop class, but I convinced him that I could do it so he let me try. I dug out every book I could find on the making of skis and I set to work. Actually at that time there was not much material available on the construction of wooden skis, so I had to make several trial and error stabs at forming and bending the wood to get the right curve and the correct surface.

When the skis were completed the next step was to make some kind of binding to fasten them to my feet. I experimented with several different designs and finally came up with one that would work. I could hardly wait for winter to try out my beauties. Ski design has come a long way since I built my first skis, but I was very proud of my efforts—and what's more they worked!

Later when business improved in Carmel and I was becoming established with my new shop, my interest in skiing returned. Through a friend, I started exploring and skiing in the Sierra Nevada Mountains. I was fairly familiar with the mountains from summer outings and weekend camping trips, but I hadn't done a lot of skiing. The kind of skiing that we were doing in those days was quite different from the modern ski resort type of activity that we know today. Ski mountaineering was what we called it and it involved climbing up the slope and when you reached the top, you literally buckled your skis to your boots (hiking) and skied to the bottom. (No fancy plastic boot in those days.) When you reached the bottom you reversed the procedure. We used seal skin climbers strapped to the bottom of the skis so that we could climb back up the slope. No fancy gondola cars for us!—simply because there weren't any!

Over the next few years my interest in this sport—some might say obsession—grew to the point where I couldn't wait for the weekends and vacations to get up into the snow. I made several trips to the ski huts in the National Parks, where we could stay over night. We would have a warm fire and a little wine and a truly wonderful experience. I was hooked.

During this period when I was doing a lot of ski mountaineering, I developed and perfected a ski binding that was a combination of cross-country and downhill. As most skiers know, the bindings are quite different in both form and function. The downhill skier must rely on his heels being bound tightly to the skis to control his ability to twist the ski and thus control the direction he is going. Also the control of his edges is critical to speed and directional adjustments. The cross-country skier, on the other hand, must have his heel free to

Demonstration at Forge in the Forest. Carmel, California, about 1962.

give the thrust and glide so critical to his forward momentum and his ability to climb up the easier grades. Ski mountaineering is a combination of the two activities so a combination binding with the properties of both or at least the capability of being changed from one function to the other was necessary. I combined the functions as best I could and was successful enough to have the binding patented. When I moved to Aspen, I sold the patent to help finance my move.

During this time, the development of skiing as a sport and as a business was growing by leaps and bounds not only in California, but across the country as well. I was becoming more and more interested in the downhill phase, and I started venturing off to ski resorts in California, Idaho, and Nevada. Of course by skiing in other areas, my circle of skiing friends grew larger and larger, and eventually we got together a group to try the fabled mountains around Aspen, Colorado. Aspen was a legend among skiers and we had to see it and experience it for ourselves.

My first visit to Aspen was in 1952. Some doctor friends, a few business associates of mine and I rented a rather large house and moved in for the week. The skiing was the best I had ever experienced. The fog and the long lift lines that had driven us out of California were non-existent in Aspen. If you wanted to ski in California, you had to drive for four hours to get to the ski area and then wait in long lift lines wasting half of your skiing time standing in line. In Aspen, the air was clear and pure, there was plenty of sunshine, and the powder snow was fantastic. Even though there was only one ski lift, the lines were short and waiting was minimal. Plus the town itself was a sight to behold.

Aspen in the fifties was a skier's heaven. The town was an old Victorian village with friendly people, quiet nights, and scenery unparalleled in this world. Ajax mountain came right into town and people were skiing down the streets. There were no traffic lights in those days and hardly any traffic. I'll never forget my first trip to Aspen, and how I hated to leave when our week was over.

On the drive back to California, a plan began to formulate in my mind that would not come to fruition until 1963, but the seed had been planted and it was just a matter of time until it came about.

Over the next ten years, I returned to Aspen to ski every chance I could, and each time I returned the feeling grew stronger—I wanted to not just visit there, I wanted to *live* there! The question was how could I afford to give up my very successful and lucrative business in Carmel and move to a town where there was an extremely small resident population and hardly any new construction? Well, as usual in life, if you want something badly enough in life, you will find a way to make it happen.

The homes that my friends and I had been renting in Aspen were privately owned and were really very nice. However, the ironwork found in the homes was generally of very poor quality. There were fire screens that would not work, fireplace hoods that would not draw, cheap hardware and railings that were really unworthy of the quality of the rest of the house. All this got me to thinking. Even though Aspen had very little building going on at that time, compared to Carmel, maybe I could build up enough business by replacing and upgrading the iron in the existing homes there. Generally the people living there could afford it, and it certainly would be a beautiful place to live and work. It just might work.

4

THE ASPEN YEARS 1963–1983

"Two Roads diverged . . ."

As I was driving back to Carmel the idea began to make more and more sense to me. Especially the part about walking out your front door to go skiing. In the end I think the atmosphere of this charming ski town was what helped me make the final decision. I am not the only person who has been charmed by the Aspen valley and the lifestyle offered there.

My first step in this cautious plan was to send some photos of my work to a friendly lodge owner that I had met on several of our trips. He showed them to the bank president (everyone knew everyone in Aspen in those days). The president said he thought there would be a market for my work there. It was not overwhelming, but then I didn't need much encouragement.

When I received the letter from my lodge owner friend, I was very optimistic. Carmel was losing a lot of its charm for me and I was ready for a new direction in my life. My marriage was breaking up as well, and my children were taking sides—life was getting very uncomfortable for me there.

I sold my shop and gave most of the money to my wife. I packed up my collection of thirty-six years of blacksmithing and headed for Aspen. Some might say I was little old for a change of direction like this, but as it turned out it was the best decision I had ever made. I was fifty-six years old and most men my age would have been looking forward to retirement, but I was ready to start over again, and I thought the future would be exciting and a good change for me. I couldn't have been more right. I have always been an incurable optimist!

It was heartbreaking giving up the Forge in the Forest and all the friends and customers that I had accumulated over the years. But the hardest was giving up my family. My children were grown and on their own now, but it still was a very difficult time for me.

On my last trip to Aspen I had purchased an old miner's house which would give me a place to land while I was getting settled and my shop built. The house combined with the two lots that I had purchased to build my shop made me somewhat of a land owner in Aspen. My ski friends thought I was crazy. They said, "You paid *that* much for *that* piece of property?" They thought I had been smart years later. I had hired a huge semi-tractor and trailer and loaded it with 33,000 pounds of equipment and supplies. I told the driver where I wanted it delivered. He had never heard of Aspen and was not real sure where Colorado was, but I gave him a map and pointed east and he was off.

With my loaded pickup truck I was soon off too. Driving across Nevada, I passed the semi and thought I was making pretty good time. I planned to have a day or two to get settled before the arrival of my equipment and work out a plan for unloading and storage while I was building my shop. As I settled in my bed that first night in Aspen my head was spinning with details of trying to make my dream come true. I knew it was not going to be easy and I hoped the skiing and the lifestyle were going to make up for loss of the established business and secure income I had in Carmel. I left behind a seat on the City Council, as I had refused to resign.

When I woke the next morning and looked out the window—there was the truck! In a moment there was a knock at the door—it was the truck driver. He wanted to know where I wanted the load dumped! So much for the two days rest and planning! I called the only person that I really knew in Aspen, my lodge owner friend, and asked for suggestions as to how I could get the truck unloaded and my equipment stored. He said as far as he knew the only fork lift in Aspen large enough for my heavier equipment belonged to the Aspen Lumber Company and that I should contact them. I did, and they were helpful in allowing me to rent their fork lift. They even sent two employees over for the day to help with the unloading! Such were the people in Aspen in those days. With my equipment and supplies sitting in my driveway and covered with a tarpaulin, I paid off the truck driver and he was soon on his way back to sunny California.

While I was finishing up closing my shop in Carmel, I had consulted a list of contractors, that I had copied from the Aspen Yellow Pages of the telephone book. Before I left, I had some plans and specifications for my shop drawn up by an architect friend of mine. He was a student of the California Gold Rush style of building, and he designed a building that would fit the Victorian style of the old mining town of Aspen. I wanted my shop to blend in as unobtrusively as possible with the Victorian architecture of Aspen.

Before I left Carmel, I sent off copies of the plans to several of the contractors I had on my list asking for bids for the job. When I arrived in Aspen, I went to the contractor's office, the lowest bidder of the three, and soon had a signed contract with a fine builder. The only condition was, I had to work for him as a carpenter's helping to get the job going—it was their busy season. I easily agreed to this, as I wanted to get started as soon as possible and I had plenty of building experience during my early years in California.

The zoning and building restrictions were not very tight in Aspen in those days, and by the end of the month I had a roof on my shop and I was ready to start. (The contractor still couldn't believe that I wanted a *dirt* floor!)

By the end of another thirty days my building was completed and I was putting the last of my tools away and opening for business. The County Tax Assessor had a little trouble assessing the value, for tax purposes, of the dirt that I had for the floor of my shop. I said, "Gentlemen, it has no value, it was there when I started to build!" That "floored" them and I won my point!

I'll never forget one of my early commissions in Aspen. One day a man came into my shop and asked for a wall decoration like one I had done for the bank. He asked me if he could watch me work and he promised not to interrupt. I thought this was unusual, but I agreed; with the stipulation that he be a good listener. He said he would be, so we started. He was very well dressed and quite reserved. He stayed with me all day and when the piece was finished, he looked me in the eye sadly and

Aspen, Colorado, 1965.

said, "I wish I could do something besides make money." I have often thought of that day and the lesson that we both learned. He learned about the joys of creating and working with ones hands, and I appreciated even more deeply what a joy my chosen profession can be.

Working in Aspen was certainly much different than working in Carmel. In Carmel, even though I was the only blacksmith, ironwork and blacksmithing were fairly commonplace, and with all the Spanish style architecture in the area there were iron gates and grilles everywhere. Especially since I had been there for almost thirty years, but in Aspen, ironwork of any kind was almost unknown and good ironwork had never been seen before by most of the people who lived there with the exception of the many fine Victorian fences that were in front of some of the older Victorian homes. I continued the tradition of the Victorian fences and through the years made hundreds of feet of them. Hardly anyone in Aspen had ever even been in a blacksmith shop. As soon as the word got around town as to what I was doing, I had all kinds of curious faces at my windows and people just dropping by to get acquainted and watch me work. There is something fascinating about watching a blacksmith at work. He is like an alchemist conjuring up shapes and forms which, in some magical way come together to make something beautiful and functional as well.

Some of the first people I met in Aspen, outside of my shop visitors, were those I met while having dinner at the famous Red Onion Restaurant. They had a "Stamm-tisch" which was a community table where anyone could sit down and order without being formally seated first. This type of arrangement was common in Europe and I was used to it from my Berlin years.

The owner was a German who liked to bring his European traditions and his foods to his socially underprivileged American cousins. He and I enjoyed exchanges in German and we liked to converse in a language that I had learned as a young man in Berlin.

The Red Onion was a gathering place for many of the permanent residents of Aspen in those days and I felt it was a good way to get acquainted and to certainly enjoy a good meal! Many of my early contacts both socially and though my business paid off in commissions. The contractor who built my shop developed into a good customer, and my contacts with architects gave me immediate standing in the building community.

While the years in Aspen were busy ones, there was still time for civic duties. I was appointed to the Planning and Zoning Commission, rose to be chairman, helped adopt the Aspen Area Master Plan, later many years on the Aspen City Council, where I continued the conservation program begun in Carmel. I pioneered the Aspen Open Space Program, a very successful effort to protect and preserve the open spaces and entry to Aspen, thus preventing the usual strip development found in so many cities.

My analysis of the situation in Aspen had proven to be correct. From the day I opened my doors there never was a dull moment. The demand for quality ironwork never ceased.

I was lucky enough to move to Aspen just before the beginning of the building boom. Aspen had been a ghost town after the mining crash in the 1890s and had never been discovered by the ski crowds as yet, but this was all about to change. When I moved to Aspen, you could have bought an acre of ground at the bottom of Little Nell for $15,000, today it would be more like 15 million!

Texans were discovering Aspen in the sixties. With the oil boom in Texas, the Texans were searching for ways to spend their money and they had their "Eyes on Aspen." You might hear the strains of "The Eyes of Texas Are Upon You" being sung in the Aspen streets at night and it certainly was true!

Before long the hills around Aspen rang with hammers. Not the picks of the miners of long ago, now it would be the carpenter's hammers that would make the hills ring. The building boom in Aspen had begun. The twang of the Texans and the Oklahoma oil men could be heard everywhere.

The Texans wanted to build condominiums. They could live in Aspen when they wanted to, but the apartment would also be available to rent out to others when they were out of town. Also there could be two, four, or six units which would be paying for the building and the owners could come and go free and make a profit. This scheme had worked well in other ski areas and would certainly work well in Aspen. From my standpoint however, it practically destroyed the identity of the small ski town, which I had grown to love and had moved here to enjoy. Such is the way of progress.

My fears of a lack of work were quickly dispelled however. I was soon being contacted for commissions in private homes, churches, condominiums, and public buildings. My reputation spread like wildfire and very quickly there was a waiting period for the products of the new Mountain Forge.

One incident I recall from one of my earlier contacts with the moneyed society that was moving into Aspen during the sixties and the seventies. My client was a Texan who had contacted me for a fire grate, a screen, and some fireplace tools for his new vacation house—as he called it. He was in the steel fabricating business in Odessa, Texas. When I finished the pieces, I called to see if it would be all right if I installed them while they were in town. The clients said it would be all right, so I took the pieces out to the house.

When I finished, the man of the house said, "Sit down, and have a beer, I'd like to talk to you for a while." I did, and we talked for about an hour about my business, his business, about my other work and some of the history of the blacksmithing profession.

At the end of the hour, I got up and said, "I'd better get back to the shop and close up." It was about five o'clock. I said "Do you want me to send the bill to your business office in Odessa, or to your address here?" I wanted him to understand that I was not doing this for the social contacts.

He said, "No don't bother, I'll just pay you now."

Well, I never pass up the opportunity to be paid on the completion of the job, so I agreed.

He said, "Just tell me what it is." The bill came to $328 which, at least to me, was quite a bit of money at that time. (That was twenty years ago.) I handed him the bill and he reached into his pocket and pulled out a roll of one hundred dollar bills which was as big as a light bulb. He then peeled off three one hundred dollar bills and laid them on the table and called over his shoulder to his wife. "Hon, got some change?" It seemed he never carried anything smaller than one hundred dollar bills! I have never forgotten that experience. That was my introduction to the new moneyed society of Aspen.

Daily I began to get requests for fireplace tools, grates, railings, gates and candlesticks. Over the years the famous and the not-so-famous came through my doors. Mortimer Adler, John Stroh, Joseph Coors, John Denver, Harold Pabst, Marty Timken (of the roller bearing family), Leon Uris, and Richard Bass (the Dallas businessman who founded Vail), all were my customers during this period. These were the scions of American industry, oil, food, clothing, furniture, Coca Cola, lumber, cereal—they all came to the Mountain Forge. They all appreciated quality and I was glad for the opportunity to provide it for them.

Another memory I have of the early days of Aspen involves skiing. At that time there was only a single chair lift to the top of the mountain which was quite slow. (This was before all the development.) We had walked over in the morning to go skiing for the day and were greeted by a long line waiting to ride the single chair lift up. Rather than stand in the long line we'd walk up there early and take our wooden skis and place them in line and stick our poles in the snow and go off to the Skier's Chalet or walk downtown for breakfast. We'd come back in about an hour and there our skis would be. Sometimes the people standing in line would move our skis along for us to keep our place in line! When we got back we would resume our place and soon would be on our way up the chair lift. Can you imaging leaving your skis in line or unguarded today with all the theft and the unfeeling people? Things have changed a lot since those glorious early days in Aspen.

In those early days, people skied on wooden skis. They were very slippery and hard to control, but that was all that we had. I remember clearly the first time I saw a pair of metal edges grafted onto a pair of wooden skis. I thought they were a grand idea and I immediately set about to make myself a pair of steel edged skis. I sent off to one of my steel supplying companies and had them send me some stainless steel that was thin and long enough to fit on the edges of my skis. I drilled and countersunk them and ground and beveled the edges and fitted them to my wooden skis. I rabbetted out the edges and fitted them on. They worked great! No more slipping sideways for me! I used those skis that way for several years and kept them sharp for much better edge control—I still have them hanging in my shop.

In the spring, the town of Aspen would be deserted. The skiers, when the snow was all used up, would move north to Jackson Hole or sometimes to Mammoth in California. The ski season would last longer there than in Aspen.

I have skied at Mammoth in June—even till mid-June. The mountain is on the east slope of the Sierra Nevada Mountains. Others would go to Mexico or even to South America to make the ski season last almost until August. In Aspen, you were lucky if there were two restaurants open by mid-April to get a bite to eat.

The business people were burned out after the ski season was over. They would say, "We've had it!" "We need a break!" and they would shut their doors and take a few months off. They would go off, as Mead Metcaff (the owner of the Crystal Palace night club and restaurant) would say, "To renew the tissues."

And then when I was on the planning commission, we couldn't get a quorum because most of the members were out of town to get away from the skiing atmosphere for a while. In the fall, we had another problem. Some of the commissioners were avid elk hunters and they would be off hunting. We couldn't get a quorum then either. The people of Aspen have certainly been an interesting group to work with—to say the least.

In those days, people placed their lifestyle first. They came to Aspen to live there as their first priority. They wanted to enjoy the country, the scenery, the recreation, and the many other opportunities that existed only there, and to hell with many of their other responsibilities. That was one of the many reasons that I was attracted to living in Aspen myself. Many of the people living in Aspen at that time had been very successful in their lives and had lived the rat race and they wanted now to slow down and enjoy the fruits of

Driveway gates. I used the threaded connection instead of pierced holes. This method is explained in detail in *The Blacksmith's Cookbook*, pages 54 and 55, 1982.

their labors, and Aspen was just the place to do that. The residents in those days wanted to live free, interesting, active, and creative lives. That attitude is almost gone now. Now all you hear is, "How much money can I make and how quick?" It is sad that that Utopian philosophy no longer exists in Aspen.

Recently, I went up to Aspen from my home in Carbondale to ski and to stay with some friends. They had rented a two bedroom apartment, the only one available, and since there were only two of them, they were paying the rate for one bedroom. I went there with the understanding of the landlord that while I was there the couple would pay the two bedroom rate. When it came time to pay the rent, the landlord said, "We can't charge you for Francis, he is an institution here!" That gives you some understanding of my reputation in Aspen even today.

While I was on the planning commission I, and many of the other members, tried to get some control on the development of this still rather small area. In fact, I tried to get the downtown area with the beautiful old Victorian homes designated as a National Historical District, this was in the mid-sixties, but we were unsuccessful at that time. Now they have done some piecemeal designation of a few parts of the city, but it would have been much easier then to designate the whole town and keep the Victorian character throughout.

It is really unfortunate that the historic district designation was not accomplished when I wanted to do it. Downtown Aspen would have looked much different than it does today. "Without vision—the people will perish," has certainly been true in this case. I had seen it happen in Carmel and I really didn't want to see this repeated—but alas, it did.

I did my best to accomplish this. It would have been so easy at the time. Other towns have done it, such as Crested Butte, but it was just not in the cards for Aspen at that time.

Thirty years ago there were millionaires who would come into town and buy one of the older homes and quietly rebuild the house and bring it up to code and it really would make a wonderful addition to the existing town atmosphere. These people didn't make a big fuss and try to throw their weight around like the developers of today.

The county commissioners of today have not seen fit to take control of the situation. There is one home—if you can call it that—in the Aspen valley which has over fifty-five thousand square feet in it! It was built by one of the Sheiks from Arabia. It has more square feet than the Aspen Airport and uses more water than the largest condominium complex! I think the growth has gotten completely out of hand and should be stopped, but the movie stars and the millionaires keep coming in and the building continues. This was one of the main reasons that I moved out of Aspen and down to Carbondale. In Carbondale we have more open space, greater views and a great deal less congestion. I love it where I live now, and I am still only about a half an hour from the slopes—if only I could find a place to park when I get there!

The character of the old Aspen is gone. It used to be "Mom and Pop" operations here where you could spend a few minutes with the owners and talk over the skiing or business in general, but now many of the businesses are just branches of large corporations and the best you can do is a sales clerk or maybe a store manager. The personal service and the small town friendliness are almost all gone. It is sad indeed.

By far the most enjoyable experience in all my years there was meeting and marrying my Portia. I had lived in Aspen for about three years and I was getting to know most of the people who were more or less permanent residents. My business was really taking off and everything was going well, but there seemed to be something missing.

In 1966, while installing a commission in Sterling, Colorado, I was blessed with meeting Portia Curlee. Our meeting was entirely by accident. I had gone there to install some fireplace equipment that I had completed at my shop in Aspen for a client. When I got to the job site I found that I didn't have the right kind of fasteners to finish the installation, so I went down to the lumber yard to try to find the correct fasteners. When I went in, behind the counter was this charming, personable lady of my age and through the course of my finding my fasteners, we became acquainted. She was interested in my profession and the job I was installing, as it was for one of her very good friends. She told me she had been to Aspen many times while taking her sons skiing there.

After the usual questions and answers it turned out that we had many mutual

Balony rail, the theme from Constitution Hall in Philadelphia, a railing made 200 years ago. 1987. Note the special winter finish.

friends and I had skied with one of her best friends from Denver. Portia suggested that we go to the White Kitchen for coffee and continue our conversation and I was most happy to comply. I had also skied with Portia's brother in Aspen and had helped one of her good friends, Barbara Brown, with a child who had been injured on the slopes. So our trails had run almost parallel for several years, but they had never really crossed until this fateful day.

At the end of our delightful morning together we exchanged phone numbers and when I had finished installing the job, I returned to Aspen with my head whirling. Before long the phones were ringing quite regularly between Aspen and Sterling. Her kids ribbed both of us into getting married because the phone bills were getting so high!

We just hit it off. It was a case of love at first sight. I proposed to her in thirty days and we were married in sixty!

Then began over twenty years of what her sons called "our fairy tale life." (See the *New Yorker* article for our honeymoon.) We traveled a lot. Portia was a long-time Colorado University football fan, and she had ten seats on the forty-five yard line, so we went to all the home games there, Portia had lived in Sterling for many years and had many friends in that area. We went to tailgate parties and parties after the games. We also went to concerts and operas in Denver and Santa Fe. And, of course, we attended many of the Aspen Music Festival Concerts. We had so many of the same interests, our lives seemed to just blend together.

My promise to Portia was to have the "best ten years of our lives together." We were fortunate in having twenty-two.

Portia and I were both mature—at least somewhat. We both had been successful in our business lives and had developed wide circles of interests and friends. We honeymooned in the Caribbean.

We purchased a condominium in Hawaii and spent some of the off-season from Aspen there. Of course during the year we traveled extensively. Our years together were full of love and life. We were like two kids fulfilling all those dreams that we had while working away at our professions. Our many years of hard work, dedication to quality and service to others had paid off and we meant to enjoy the time we had together to the fullest. I felt that my life was beginning all over—at sixty! It was wonderful.

Portia's mother and family were in the top echelon in the field of medicine. Dr. Portia, her mother, had a fascinating life. She had been born into an old Southern family and had met and married a Russian agronomist. They moved to Russia where my Portia was born. When the Russian Revolution came along in 1917, they were forced to flee with Portia being carried out in her mother's arms. It was a fascinating story and made a wonderful book entitled *Dr. Portia*.

Portia was determined to enjoy my lifestyle to the fullest, as I was hers. She had never been a skier and her trips to the slopes were to take her four sons for their lessons and let them learn to ski. She had waited impatiently at the foot of the slopes. But now she knew that if she was going to be with me, she was going to have to learn to ski herself. Bravely, at age fifty, she took lessons and gradually became a very good skier. We skied together for most of our married life. This attitude of sharing each others interests was one we both believed in strongly. I still think the secret of a happy marriage is based on this concept: *Enjoying activities together and not being jealous of the time a spouse spends doing something they enjoy gives both people a stronger feeling of togetherness. Doing things together is the secret.*

In 1976, Portia and I were invited to the Artist-Blacksmith Association of North America Conference, in Carbondale, Illinois. Throughout my whole career I had worked alone in my shop and had never attended shows or crafts fairs to present my work to the public. A friend of mine had told me of seeing a news clip on the formation of a National Association of Blacksmiths on television and he wondered if I had heard of it. I had not, but I was immediately interested. When Portia and I received our invitations to attend the first conference we were delighted and made immediate preparations to go. I had no idea at that time as to how much this meeting was going to change our lives in the next twenty years.

It was at the conference in Carbondale that I received the call, as they refer to it in religious circles. I met many young smiths with a burning desire to bring back the art of blacksmithing and now I had a mission in life. The young smiths had no one to turn to for information and skills. It was there that I realized that *I* was a link between the heyday of wrought ironwork in the early years of this century, and the present renaissance of blacksmithing. I had the year with Samuel Yellin and I had the training in Germany which had given me a solid background in the traditional techniques. Now it was up to me to not let these techniques and knowledge die. I immediately let it be known that I was willing to put on a workshop wherever and whenever we could get a group of smiths together. Portia and I started what we called our road show and it has lasted up to the present.

We were flooded with calls and requests and we put together a series of workshop tours, which eventually numbered over one hundred and fifty! We presented work shops in twenty-six states from Alabama to Alaska. Portia loved it. She was my greatest helper and admirer. She was the travel agent, correspondent, recording secretary, and anything and everything that was needed to make my life run smoothly and efficiently.

A flower gate for a private residence in Aspen, Colorado. 1984.

We were a great team.

After several summers of touring and putting on workshops, I decided to write a book of instructions and techniques that I could leave with the smiths when the workshops were over. The book would serve as a permanent source that the smiths could work with all year round. The result was *The Blacksmith's Cookbook,* which Portia helped me write. The format was her idea. The book has proven to be a very successful educational tool for smiths who want to expand their skills and techniques. I have donated all the proceeds from the sale of these books to the Francis Whitaker Blacksmith's Educational Foundation which provides funds for workshops, scholarships, and other blacksmith educational activities. This idea has worked very well and it will continue. The modern blacksmith works basically alone and it is sometimes difficult for him to get new ideas and techniques, or to find out about the old ones. I am sure my workshops and my *Cookbook* have opened the eyes of many beginning and experienced smiths to the possibilities with the forge and anvil.

Portia was full of laughter and practical jokes. She was the perfect complement to my sober and serious side. She always made sure I didn't get too serious and that our lives would be lively and happy. She was also an excellent accountant and did all the bookkeeping for our business, The Mountain Forge. Our workshops took us to many places we would not have otherwise gone to, we were always among friends.

Perhaps our greatest trip, and the last major one, was a vacation trip that we took to New Zealand. We decided to stay a month, as we wanted to meet the real New Zealanders not the ones on the postcards. We made arrangements to stay at bed and breakfast homes.

At Rotorua, we stayed with president of the local Bank of New Zealand. He and his wife took people in just to meet them and they were especially interested in Americans. Their house was heated with steam from the underground volcanic activity. A heated swimming pool, sauna, outdoor cooker were all heated by the heat from the volcano. To tell the truth, all that volcanic activity made me a little nervous, though it never bothered our banker friend and his wife. I guess I had lived in California with the San Andreas fault as my neighbor too long.

At one stop there was not a bed and breakfast home available, but we were told we could get a "farm stay." We didn't know what this was, but we decided to find out. Were we surprised and delighted! A winding road led us up to a beautifully built old Victorian home, sitting on top of a hill. It was like a setting for a movie. The house had fourteen foot ceilings, arched hallways, a four-car garage, a swimming pool—some farm! The owners invited people into their home just to meet tourists and they too were especially interested in meeting Americans. When they found out that I was a blacksmith, interest was high.

On the south island, we traveled by train, boat, and bus just for a change and to meet more people. Of course, as we traveled anywhere we were interested in the ironwork and in New Zealand it was outstanding. The smiths there were from Scotland, for the most part, and they had access to good coal and iron on the islands. The result was that they produced truly remarkable work. We stopped at a little known small city, Timaru, which had barely been mentioned in the guidebooks, and the ironwork was magnificent. Most noticeable, including the fine cast iron balconies and railings, were the colors! The works were painted white, orange, green, blue, and purple, anything but black. Portia and I took hundreds of slides and the work gave me more inspiration for my own projects.

We stopped at the shops of a couple of working blacksmiths on the islands and found them both doing excellent work. It was very gratifying to see such fine work and know that I was not alone in keeping the art of blacksmithing alive.

As we traveled around New Zealand we had a little trouble with the Scottish brogue, the terminology, and with the Maori accent, but all-in-all it made the trip more interesting and enjoyable. Although we were sometimes left scratching our heads, we had never met a more friendly, open people.

One of the delights in the early days in Aspen was spring skiing. My ski friends from California wanted to know how long the ski season lasted. I started counting the months and when I got up to twenty-two, I knew that we could ski all year 'round. The lifts in Aspen closed in April and all of the out-of-town skiers went home. But this was just the beginning of our spring skiing season which, we thought, was the best of all. We climbed Aspen Mountain and skied on the spring snow, which was delightful.

By June, Independence Pass was open and we moved over the mountain to ski the many runs there. Portia would drive me to the top, and turn around and go back down and watch me with her field glasses as I skied down. I have skied a run called The Fourth of July Bowl on that exact date. August it was up to Castle Peak, a dead glacier called Montezuma Basin, which had a small portable cable lift. I earned a lifetime pass for helping put it up, but now it is gone. May would also be the month to climb Mt. Sopris carrying our skis to the top and making a terrific run down. I often think of those days as I look out my front window at Mt. Sopris today.

The early snows on Independence Pass would account for September and October, and then we would be back to the lifts in Aspen by Thanksgiving. Skiing has been an obsession with me since I first started

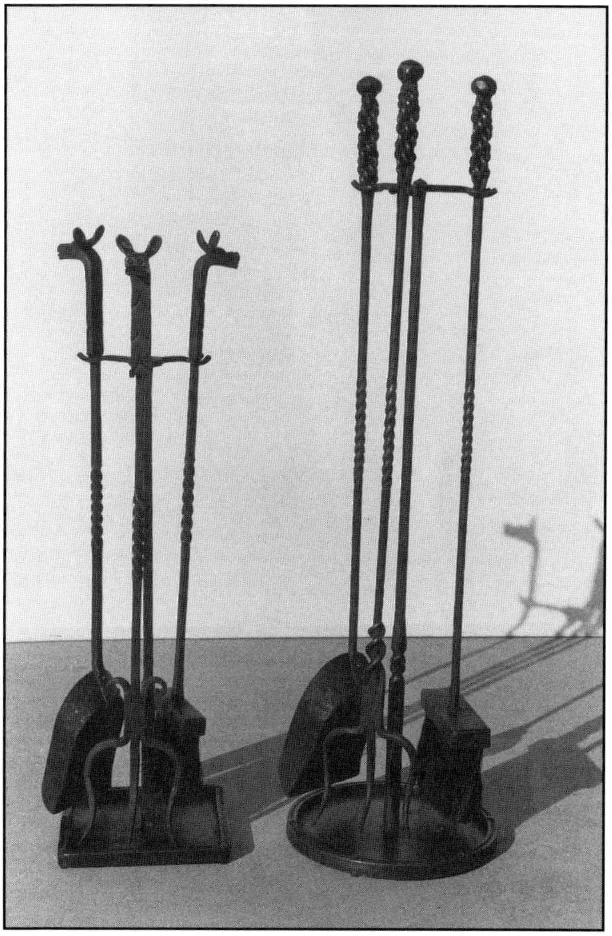

Two sets of fireplace tools. 1982.

in California many years ago. However I believe it has been the outdoor air and exercise that has kept me young and active all these years. There is no substitute for regular exercise and a spirit of adventure. People today are just beginning to realize the benefits of such a life both physically and mentally.

My business at The Mountain Forge was steadily increasing, so I started looking for someone to come, work with me, buy my stock in the corporation and eventually take over completely. For ten years, I had a series of blacksmiths come in, work for a while and then leave. One or two lasted for a year, but the answer was always the same. They all wanted the benefits, but none of the responsibility. Increased federal and state regulations and added-on fringe benefits soon took all the fun out of running your own small business. I can see very readily how so many small businesses fail in the first one or two years. There is very little incentive to start a small business today with all the red tape and regulations. Forty years ago the small businessman was the backbone of America—alas today it is not the case.

Once one of my former employees deliberately picked a fight with me and quit. He went to the unemployment office and filed for unemployment benefits claiming he had been fired. I discovered that he had already set up a blacksmith shop in California, yet he expected to receive unemployment benefits from me! That was the end of my having full-time employees. I went back to working alone, my overhead and bookkeeping costs went way down, I was much happier and enjoyed my work immensely.

My largest commission in Aspen was all the work for the restoration of a large and very fine Victorian home. I made

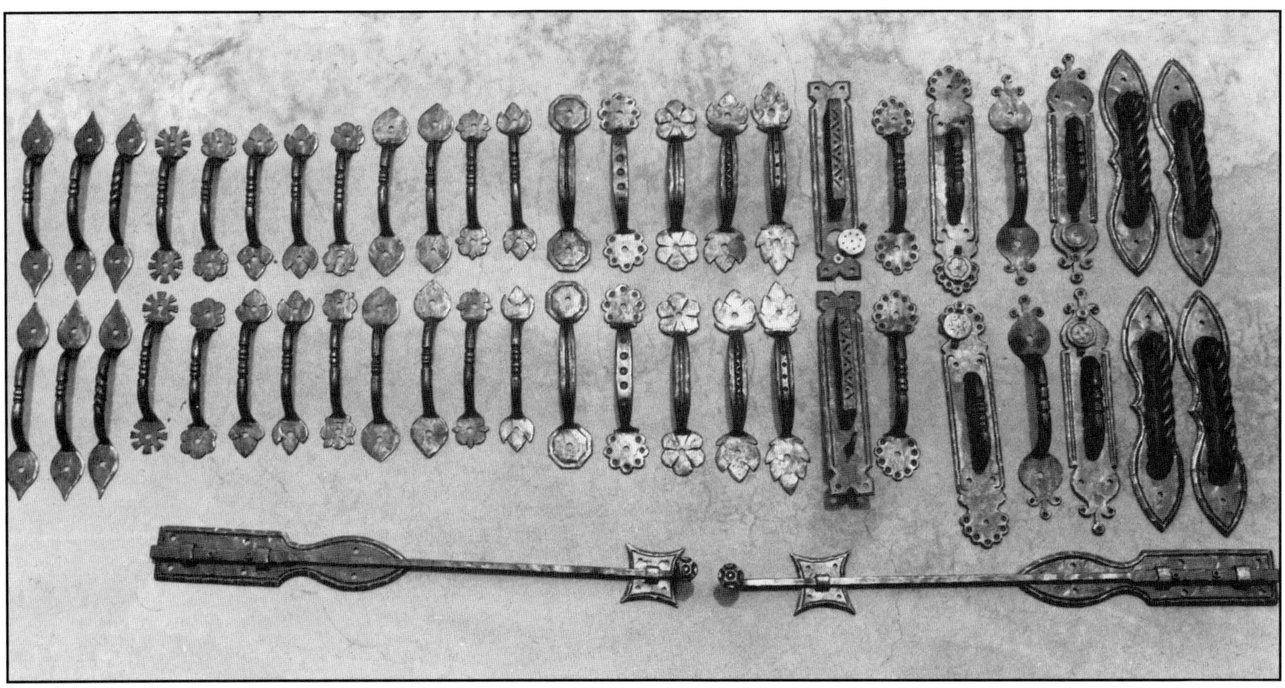

Part of fifty-three pair of door handles. The client wanted each pair with a different pattern. A fun job. Brekenridge, Colorado, 1984.

gates, railings, balcony railings, finials, fireplace tools, and fences. The whole home was restored and it turned out beautifully. I especially liked the idea of creating pieces for the entire house as some of the themes I was able to carry throughout the property. It was a very satisfying job when completed.

Outside of Aspen, my largest commission was for a new home in Breckenridge, Colorado. I made one hundred and six door pulls for that one house and it stretched my imagination to the fullest to come up with fifty-three different designs. I also made more railings, many fireplace furnishings—complete with screens, tools, and andirons. I even made some of the light fixtures. The final touches were the large entry and driveway gates. It was an immense project, but quite rewarding. I also have the satisfaction of knowing that my work will be viewed and appreciated long after I am gone.

The most difficult commission of my entire career was a balcony railing in Aspen. The railing was elaborately scrolled and curved both in and out. I included it as the last picture in my book The Blacksmith's Cookbook. It was a very complicated design and I guess it both fascinated and frustrated me. This is true of blacksmithing in general, but the railing turned out to be much more complicated than I thought. Finished, it was very beautiful, but it was a real struggle getting there. In the end, I was very satisfied with it.

My farthest commissions from home were a fire screen and fireplace tools for a home in England and a set of fireplace tools for a home in Australia. So I guess you could say my work has international appeal.

Generally, over the years, I had little trouble collecting for the work that I did. Of course, one advantage that I had was that the clientele I was doing the work for were generally in the upper income class. Also I would require a deposit if the clients were unknown to me. What I usually heard was, "Mr. Whitaker, you do fine work, and we would like you to make this commission for us. When you finish, send us a bill." It takes many years of fine work to get to that point, but it is the most sincere compliment a craftsman can receive.

The accidental discovery that Portia had cancer brought about a complete change in my life. Everything that loving care and medical science could do could not save her. A struggle of a year and a half and then the end came. Needless to say, it shattered my world. I still miss her every day. After Portia's death, Aspen became increasingly uncomfortable for me. The loss of Portia sent me searching for another place to live and carry on my work. There were just too many memories around Aspen and I wanted to take a different direction.

My ten year search for someone to take over my business had ended in failure. I just couldn't find a worthy successor. So I decided to donate my shop, tools, and materials to an appreciative institution. I contacted several colleges and universities in Colorado and other states, but the interest was minimal. If I was going to donate over one hundred thousand dollars worth of tools and equipment, I wanted it to go to someone who would appreciate it!

Finally through another blacksmith friend, I found Colorado Rocky Mountain School and received immediate interest in my shop, equipment, *and me!* The school is in Carbondale, Colorado just thirty miles from Aspen, but in a different world. The merger has turned out to be an ideal situation for both of us. I have an excellent place to work, conduct workshops, train students and apprentices, and a wonderful home.

I donated the entire inventory and equipment of The Mountain Forge and helped raise enough money to build a fine six station log blacksmith shop. I feel I have finally found a good home for the Forge and, not incidentally, for me, but more about this in the next chapter.

Balcony rail. My most difficulty commission in 60 years. All parts, mostly of 1/2" square, forged to a stretched-out pattern, then bent, section by section in a heavy duty brake with the jaws set well back so no marks showed. Aspen, 1984.

5

THE MOVE TO CARBONDALE

My gradual disenchantment with the changes in Aspen, my Aspen, reached the frustration point when I lost my Portia. I was very unhappy with the way Aspen was going and I just wanted out. Out where I could live comfortably and continue my work, but where I would be away from the changes that I thought were destroying the lifestyle of the town I had grown to love.

After my unsuccessful long search for a successor and my frustrations in finding an institution that would covet and use my life long collection, I was getting discouraged in my quest to find a home for The Mountain Forge.

One day I was discussing my dilemma with Will Perry, a local artist blacksmith, he suggested approaching the local Colorado Rocky Mountain School with the idea of locating my shop on their extensive campus. The school is a preparatory boarding school for high school students which offers an extensive curriculum including an emphasis in the arts and an outdoor education program that is second to none.

I had known Will for many years since back in the days when he was a student at Colorado Rocky Mountain School and he had come with his class to The Mountain Forge to see a demonstration that I had put on for the students. Will had been intrigued at what he had seen and immediately set out to see what he could do with the forge and anvil. He became hooked, as so many others have, and borrowed an old forge and anvil from one of his father's ranch houses and started in. I called on Will several times through the years to help me on large commissions and we had become friends. I sometimes wonder how many hundreds of individuals I have started into the field of blacksmithing with one of my demonstrations. I am sure the number is staggering.

On the strength of Will Perry's suggestion and the fact that I was at my wits end, I approached the headmaster with my proposition. He was very receptive to say the least. I gave him my resume, explained my motive and we started making plans.

The Colorado Rocky Mountain School started on an old ranch site in 1953 and some of the buildings are of the original log as were many of the early ranches of this area. So we agreed that my shop should be log also. Of course this idea suited me just fine. There had been log blacksmith shops in this country since 1620.

The next step was to raise funds to build the building and finance my move. I was designated a beautiful spot right in the heart of the campus and later a plot behind the shop was leased to me upon which I could build my home. It is really a wonderful set up.

Terry Lee was extremely helpful in organizing the raising of funds. A large part of the money came from my friends in the blacksmith world, some from foundations and some from friends of the school.

My status and title is that of lifetime resident smith. I keep the shop in order, teach a few volunteer students to make things for the school. So far we have made bike racks, railings, chandeliers, hat and coat racks, and some other smaller projects.

The finished building contains a very large smithy with work stations for six smiths each with his own forge. The other end of the building contains a very complete pottery workshop with a large outside kiln behind the building. Between the shops is the Portia Room where hang some of the beautiful quilts created by my late wife. This room also contains a library of blacksmithing books that I have collected over the years as well as other displays. All-in-all it is a very complete facility and it also has been the site of two blacksmith workshops so far and I hope of many more.

An Aspen architect, Harry Teague, designed the building. The prime contractor was Floyd Diemoz. However, a great deal of the actual work was done by adult volunteers and my friends. They came in and put up all the log walls in one long work weekend. The logs came from the Flat Tops area, where some forty years ago an infestation of the spruce beetle had killed all the trees on this high mesa. They had been left standing all this time, until an enterprising sawmill operator got permission to harvest them. I like to think they were waiting there for me to decide where and when I would use them. I had seen many stands such as this when I was skiing in the high country and they seemed to be just waiting for someone to put them to use.

The logs were cut into eight by eight timbers, double tongue and grooved in the mill and they turned out as straight as a die. They had stood so long in the dry mountain air they were air cured and we have hardly had a twist or a crack.

Putting up the building was like laying out a child's Lincoln log set. The corners were notched and windows cut in and then the logs were spiked together. The walls went up very quickly. We used red aluminum sheets for the roof covering so that the snow will slide right off. There is quite a heavy snowfall in our valley in the winter and snow build up can be a problem. With a good pitch and the metal cover on the roof there is virtually no snow build up and this kind of roof will almost last forever.

Once the shell and the roof were in place we started finishing the inside. Large ventilators were necessary to carry out the smoke from the forges and I combined the hoods in pairs giving a total of three stacks. Included in the plan was plenty of storage for my large stock of materials and equipment. The Rocky Mountain Smiths put on a work weekend to move all my tools, forges, anvils, and special equipment into the new shop. The students at Colorado Rocky Mountain School were a great help with the finish work to the inside of the building. They did a lot of the carpentry, wiring, sheet rocking, as well as helping to put the roof on. It certainly was a grand day when I first built a fire in my old familiar forge and struck my old worn anvil to announce that I was back in business once again.

So many of my tools have been with me so long they are almost like old friends. My Little Giant trip hammer is still tripping merrily away after sixty years of almost daily use. My Swedish steel anvil is a little worn at the edges but has never been abused and looks almost as good as new. It's ironic that a man's body ages and wears out, but if his tools are of good quality and are oiled and well taken care of, they will outlast him.

Once my shop was set up and in operation, I opened the doors to the students on campus to come in and try their hand at the forge and anvil. The response has been very encouraging. I began teaching one class session per week and have had the forge open, when I was working there, for students to drop in and work. I have had over fifty students come to the forge and try their hand. Since they are all volunteers for extra-curricular work, their response had been very encouraging to me.

My agreement with the school is that I may have as many outside students as I wish as long as they do not interfere with the school use of the facility. This hasn't been a problem as my students are usually quite advanced and make excellent examples for the students just beginning. I have had two long-time students, one for the full school year, another for three months. I have had several short-time students who have come for refreshers or practicing professionals who want to raise the level of their skills.

In 1990 we held our first major Blacksmith's conference in Colorado at my facility and the turn out was excellent. The conference was sponsored by the Rocky Mountain Smith's which is affiliated with the Artist-Blacksmith's of North America. ABANA, the national organization, was founded in March of 1973 at a gathering of 20 smiths in Lumpkin, Georgia. Now there are twenty-nine chapters in twenty-nine states and over three thousand members.

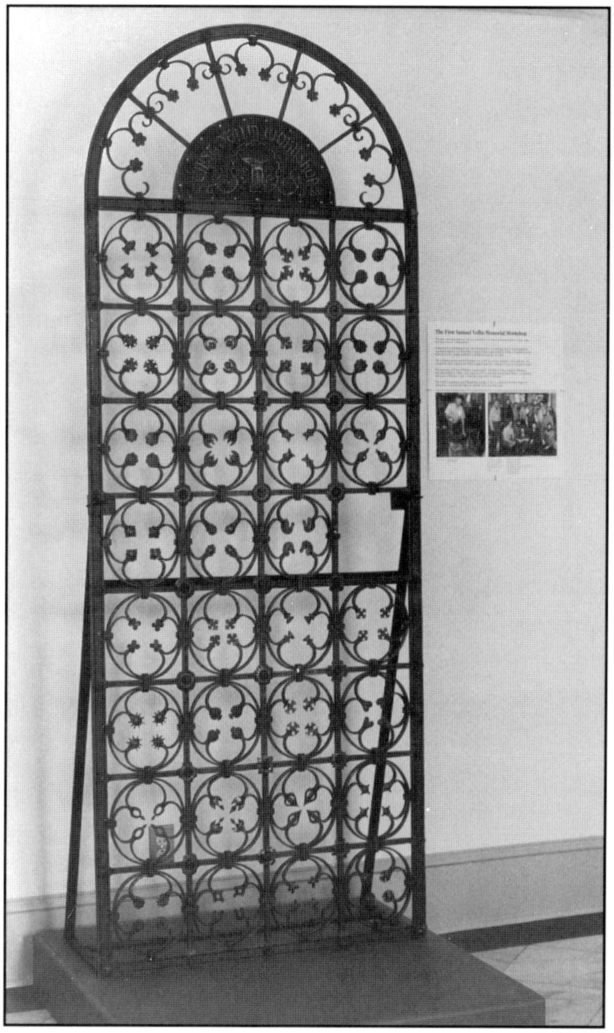

The first Yellin Memorial Workshop. This is where we worked out the forging of a quatrefoil from one piece, doubling back and forge welding the stock to gain the mass needed for the foil. Philadelphia, 1983.

ABANA published *The Anvil's Ring,* a fine quarterly, and has national conferences on a regular schedule. The greatest advantage of having a national organization is that now the information and techniques can be made available to everyone and the old tradition of "trade secrets" can be eliminated. This alone has brought about a tremendous change in the renaissance of artist blacksmithing.

I was very pleased to have our first state conference at my new shop and show what a great learning experience a conference of this type can be with the right kind of setting and equipment.

This year (1992) we held our second conference here at the Colorado Rocky Mountain School and there were over eighty smiths in attendance from all over the United States. The forges were blazing and the anvils ringing for three days and nights and a great deal of knowledge was exchanged. It is very gratifying to me to see the interest that has grown in Blacksmithing in recent years and to know that now I have a home and a wonderful shop where these traditions can be passed on to others.

I had lived in an apartment for two years, but wanted a home on campus. An architect who had done some work for the school completed the final blue prints and soon the contracts were let and the building began. I really didn't want a large house which would be a burden for me to take care of so we agreed on a rather simple chalet with several touches that took advantage of the site and the view.

I designed a large round window in the living room through which I could view Mt. Sopris and the many changes that it goes through in the various seasons of the year and even the hours of the day. I have climbed that mountain so many times I feel like I know every rock on it and it is a great joy to me to look out at it whenever I want to.

I have two bedrooms in my home and the guest bedroom is separated by the main part of the house so that my guests may have privacy as well as comfort. Eight foot sliding glass doors lead onto a large deck from the living room and each of the bedrooms. The deck runs the full length of the south side of the house and provides a beautiful spot to enjoy the sunshine and the views of the valley and surrounding mountains. My efficiency kitchen, as I call it, gives me plenty of room for the moderate cooking that I do and is still easy to keep clean and is quite functional.

The outside of the house has log siding allowing it to blend in with the shop and the other older buildings of the campus, and there is a small lake across the driveway which provides scenic beauty as well as the irrigation water for the grounds. The Crystal River is only a few steps down my driveway and I can take my walks along its turbulence whenever I choose. All-in-all, the setting for my home is about as scenic and beautiful as anyone could want. We have very little traffic as all the students live right here on campus so there is hardly any noise or pollution. I am at peace and quite satisfied with my home and my shop and have the honor of being the lifetime resident blacksmith for the Rocky Mountain School.

I am now in a position to devote much of my time and energy to the preservation of the making of traditional ornamental ironwork. Each year, which I tell myself will be the last, I am off again traveling and teaching. In more recent years, I have specialized in master classes, working with six to ten smiths on a large architectural project. Starting with a scale drawing and moving to a full-size layout, we then complete the necessary steps to the final assembly and installation. This type of experience is just not available to most working smiths and yet it is very necessary if our traditions are to continue. Working—hands on—with a few students, my knowledge can be passed on very effectively. At the age of eighty-six, I can look back on a record of mastering a stubborn material over a span of more than seventy years.

6

LIFE WITH PORTIA

I am sure that if I hadn't met and married Portia Curlee twenty-six years ago that the later years of my life would have been much different.

Up to that time I had focused on making a living, improving my life, raising my kids, and improving my techniques and style. Portia encouraged me to look out and go out and begin to share what I had learned with others.

My first contact with the Artists-Blacksmiths Association of North America (ABANA) really opened my eyes to the need for and exchange of information on techniques that had almost been lost in America. Portia was the perfect complement to my somewhat introverted personality and as a team we began an over twenty year odyssey to spread the word that blacksmithing was indeed alive and well.

I would like to dedicate this chapter to the memory of Portia Curlee. Without her enthusiasm, encouragement, management skills, love, and total dedication I would never have been able to accomplish the things that I have.

Life with Portia was never dull! I had lived alone for approximately five years and as a bachelor had never been interested much in the social life in Aspen. I had my friends and business associates and I had my work and that was most of my life at that time. My marriage to Portia changed all that.

Portia was a very outgoing person to say the least. She had been in the business world for most of her adult life and had made many, many friends. Also she had maintained a close relationship with the University of Colorado at Boulder and had many strong ties with the faculty there. And she loved to travel and see how other people lived. She was not a "ski nut" as I was as yet, but she began skiing and joined me on many a run.

We bought a house in Maui, Hawaii, and spent some of the off-season there. From my rather sheltered life, I suddenly was thrust into a flurry of activity, new friends, travel, conferences, workshops, and seminars, all with my friend and wife at my side encouraging and supporting. One summer we crossed the country six times presenting seminars and workshops. Portia loved it. We were treated like royalty and met thousands of people that we never would have met. It was a wonderful experience for both of us and we had the time of our lives. Portia's family kidded us and called us the fairy tale vagabonds, as we were always off to somewhere.

During the days when I was courting Portia, an event occurred that had a profound effect on both our lives. I was visiting in Sterling, Portia's home town, and coming into the house from the garden, I heard her son John say, "Mother, you are acting just like a kitten when Francis is around." At that moment, I put my head around the corner and said, "Meow." We all burst into laughter, but we were both transformed. Portia became the "Sterling Kitten" and I became "The Aspen Tom Cat." The license plates on our cars changed. Portia's was "MEOW K," for kitten, of course, and her new skis were engraved, the Sterling Kitten. My license became "MEOW 1," which meant that I was the number one Cat in the household.

That little word became so important in our lives. We were both mature and independent and had made our own lives alone for some time, so the blending took some adjusting on both our parts. Portia had single-handedly raised her four boys, and had put most of them through college. I had concentrated on reaching the top of the blacksmith world. Our independent personalities were sometimes in conflict, but that one word, "Meow," spoken in many ways, would always bring a laugh and understanding between us. Sometimes soft, gentle, pleading or strident, it worked.

We both felt that everyone should have a "Meow" word that would break the ice. It was a magic word for us and made our lives happier.

Always the prankster, sometimes spontaneous, sometimes calculated, Portia was hell-bent on making life interesting and she certainly did! One summer day we were up at the Maroon Bells, a beautiful pair of peaks with a crystal blue lake at the base. A tour bus pulled into the parking lot and the passengers unloaded and all lined up for a group photo with this spectacular mountain scene. Portia, quite uninvited, joined the group and beckoned me to follow, which I did. The poor photographer did not know everyone in the group, so we were recorded for posterity, uninvited and unidentified, I often wonder what the people in the group thought when they saw the unidentified faces in the photo, Ghosts? Aliens: Such was the subtle, or not so subtle humor of my Portia.

When her son Charlie was loaned our Aspen home for his honeymoon, all the light fixtures were outfitted with flashers, the bed shortsheeted. A true mother's touch.

I asked Portia's sons to contribute some of their memories of their mother to this chapter and the following is from her son Johnny.

"From the day Francis met Portia over coffee at the White Kitchen on Main Street in Sterling, Colorado, he courted her with such devotion that the romantics of old sat up and took notice . . . sharing flowers, poems, dinners, fine wine, walks in the high mountain meadows . . . He treated her like a queen, and she in-kind, treated him like a king. Their courtship and their love together had a magical quality that glistened in their eyes and echoed in their laughter. Francis was nearing age sixty at the time, and Portia was ready to send the last of her four boys off to college. Nearing fifty, Portia was looking at an empty nest. The Lord often works in mysterious ways, and so it was with Francis and Portia.

They were both rejuvenated and found a new and exciting life opening before them. (Why, the way they carried on, it was enough to make young lovers blush!) They were meant for each other.

"One thing, however, that Francis had not counted on as part of the bargain was this thing . . . this odd form of humor . . . let's call it the Curlee humor. Perhaps, the love of pranks and practical jokes was cute in Portia, but, Ye Gads, her boys have inherited it! Francis had to realize this on their wedding day.

"It was a simple and yet beautiful wedding at the First Presbyterian Church in Sterling. Francis was accompanied by Jerry Brown, his best man. The wedding went well and it was a fine reception. Jerry gave Portia and Francis a ride to the house where they were to make their get-away in Portia's '63 Chevy Impala. Everything had proceeded so smoothly. Now, after all, this was a wedding and the pranks Portia had been involved in in her lifetime would require a book alone. Contrary to what one might expect there were no tin cans, no 'Just Married' banners, no balloons . . . Portia and Francis had seen to that. The car was safely locked in the garage. No, not the Curlee's garage . . . The Weakley's garage which was across the street.

"What they did not foresee was just how persistent the 'Curlee humor' can be. Johnny and Charlie knew the chips were down. After all, how many times would they get a chance to marry off their mother and give her a send off that would be fitting to the original prankster?

"They sweet-talked Catherine Weakley, an old friend of Portia's and familiar with her prankster ways, into letting them into the garage. The boys jacked up the car and blocked it so that the rear wheels were just a fraction of an inch off the garage floor. Catherine, not wanting to be a totally innocent culprit, brought the rice to garnish the visors, suit cases, and the rest of the car.

"The moment came for the great get-away. They went to the Weakley's garage amidst the joy and hubbub of family and friends. Smugly, they eased into the Chevy,

Gate for a church in Golden, Colorado. I took the cold lifeless architects drawing and translated it into my own style. They wanted the cross and the trinity. The cross was simple, the trefoil was my first use of this motif. Note how the twists follow the circular top.

Francis at the wheel. They had backed the car in so that they could make a speedy get-away. (Portia didn't trust any of us kids after all the tricks she had played on us.) Key in. We have ignition. Shift confidently into drive. Gentle on the accelerator. Nothing. More on the accelerator. Still nothing. Step on the accelerator and the engine racing, we do not have lift off! We have laughter!

"That was one for Portia, the prankster, and her beloved Francis and one for the road.

"Well, they did finally get on the road and an enchanted road it was! The journey took this rare and precious couple along alpine streams and meadows adorned with flowers, across the ocean to the condo at Maui and the eternal sound of the surf, down under to New Zealand, to workshops for Francis from coast to coast. They started in Francis' small home in downtown Aspen and moved to the 'palace' on Snow Bunny Lane. They gardened, they watched the birds, they skied, and they entertained family and friends together. They celebrated life and life celebrated them. Close your eyes, and look closely. There they are on Independence Pass . . . the Tom Cat and the Kitten . . . the champagne is open and in the glass. Hear them meow to each other and laugh. They toast the beauty of the flowers, the crisp mountain air, the sun, and the majesty of the Rockies. They toast each other . . . Portia and Francis."

[Contributed by Johnny Curlee.]

"Being used to a more sober approach to life the merging with the Curlee family certainly gave a new dimension to my life and led to many unexpected 'surprises,' all of them pleasant."

[The following letter was submitted by Portia's oldest son, James A. Curlee, of South Bend, Indiana, April 5, 1993.]

"Dear Francis,

"Today Mother is on my mind in a special way, of course as it's the anniversary of her birthday. I am also thinking of you and I'd like to share some of my thoughts and feelings about mother with you.

"Mother loved to celebrate—birthdays, holidays, seasons—she was a notorious April-fooler. I imagine she 'got' you too as she got all of us. I cherish the way you and she were able to *celebrate* life together with the love, joy, and vibrancy that I witnessed. I am sorry for you, for all of us that she isn't here to celebrate with us now.

"The Christmas star she quilted for us is hanging over my fireplace in place of a favorite watercolor of hers. It's a special memento I just haven't wanted to put away yet.

"I'm so glad now that I am enjoying it today even though it is spring and almost Easter. Soon I plan to make some of her Russian Easter cheese on Saturday with Heather.

"Thank you for your part in encouraging Mother's experiences of creativity in your lives together, especially the quilting, the painting, and the writing. All three were good for her and for us—then and now.

"As a child, I remember some of her college-era art. Later she sewed some, especially some wonderful Halloween costumes for us kids. She even tried painting—up in the mountains, with us four young boys running around, while Daddy fished—not a very conducive environment for the budding artist, I'm afraid. And then, in the year my Dad was ill and dying, she worked with Wayne, there didn't seem to be enough time or energy for everything and everybody. Thank you for the love, energy and support you brought to her life. I am honored to hold some of the results in my heart and to enjoy them on my walls.

"I pray this finds you in good health, active, at peace, and yes, a little feisty.

"With love, Jim."

[Contributed by Portia's son, Paul G. Curlee, M.D.]

"Whence my BEING? Yes I come from Mother. Mother what? Earth? River? Perhaps I came from some yet unknown womb place. I do recall that beginning. Strange it is that I should know . . . I mean *really know.* Not just some faint recollection of what parents remembered to tell later on. I really KNOW that earliest time. 'How do I know?' I frequently ask myself. And then the answer comes, a soft musty colored feeling. I am aware of a subdued orange-colored glow with reddish blushes intermingled and an occasional lighter yellow tone, contrasting the darker orange. I'm not sure if I'm *intra utero* or outside with underdeveloped retina. It doesn't matter. The feeling of total effortlessness is pleasant. It just is.

"Later, time span unknown, but about 1–2 months I feel water, hear sounds, sense smells. Mother's voice coos gently to my face, water laps up upon my neck and chin. The water's tickling warmth enters ear canals and my heat is lowered more deeply. A cradling, cupped hand supports my neck and head. Feeling bliss, familiar comfort, I languish in mother's love. The scent of rubberized canvas marks the time (WWII era). This smell is not completely pleasant, but is comfortable, signifying familiar, soft surrounds. Smells move to touch and it is of a warm soft milk giving breast. Again surrounded by a golden aura, I sense, feel, imbibe the fullest essence of Mother. Ah, such nurturing! Bliss!

"Apart from this genesis, I don't recall much until toddling along. The memories, however, are vast. What is it that marks their uniqueness? A compendium of thoughts, actions, feelings, forces surge in my heart and mind. The constant fiber threading this history into meaning is the most precious of Portia—'Mother is at

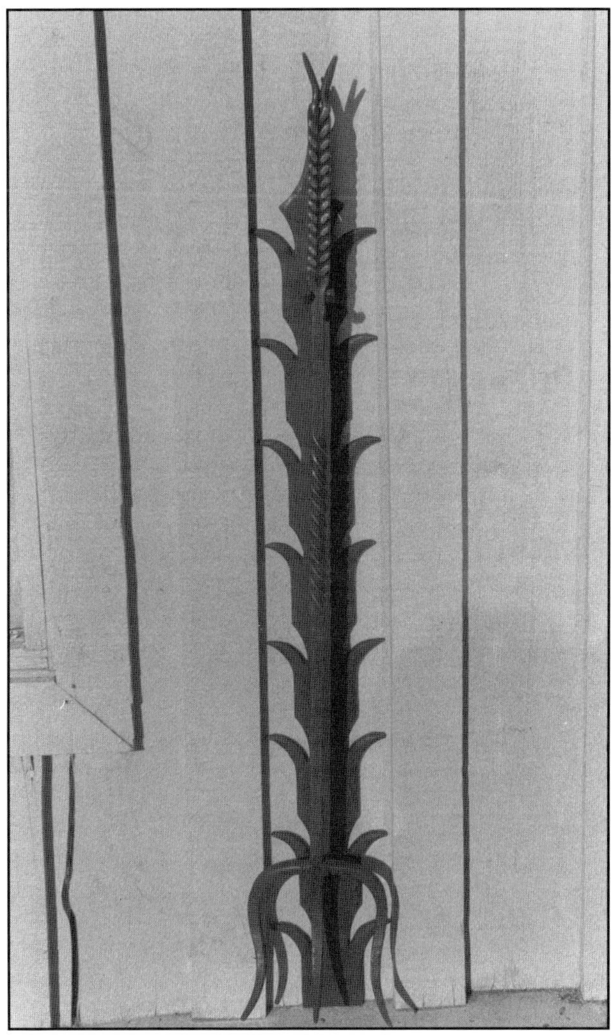

Wheat straw fireplace tool and holder. 1985.

home.' Home with me, home with all of us, and home with self are the descriptions of her. A total mother? Yes and there is more. The caring hand that toweled a feverish head also tied a scratchy woolen ear-flapped cap to keep me warm in snow play, a smile albeit stern would admonish my destruction of the Giacomine's raspberries, but also signify enchantment with her own mischief. She cared for my spirit within the structure of church, the beauty of flower and garden and the nature of love. The physical was cared for with tables of food and marvelously learned cooking. Mother Portia generated intrigue with household chores and responsibilities by modeling example, giving me skill and knowledge to make things and repair them. Through school and athletics, she was always there willing to participate, encourage and at times correct errant direction. I can't honestly recall a time she was 'not at home' for me.

"There were times when I recall differences of attitude, opinion, or desires for change. I never knew dislike or loathing as some seem to harbor for their parents. How could I do less than honor her? Contemporary cliches like 'unconditional love' are lauded about as aspiration for attainment. In Portia, this was a given, a life quality of Being, that simply was. In her it took no intellectual contrivance to develop caring and love.

"These attributes and innumerable others came effortlessly and genuinely. That spirit, that richness of loving were inculcated to me with a total caring and support. 'Thank you' rings hollow in efforts to show gratitude.

"The foregoing must not be construed as Sainthood for this mortal mother. A host of idiosyncrasies, fetishes, and even bizarre fascination with rambling details would certainly convey the notion that 'this lady is a character.' Be it participating in ruthless Halloween pranks or tricking us beyond belief with April Fool's Day shenanigans, she was always catching me off guard. With each novelty came an understanding of flexibility and the ingredients of 'good times.' Not only was Mother always up for fun, but she gerrymandered the world around her into wonderful activities. Such activities range from picking weeds (with the promise of swimming afterwards), to skiing (all ski-bums befriended and invited in for Happy Hour), to waterskiing and picnicking (she set records for trolling underwater on a ski rope), and especially just being together with a puzzle, a cup of coffee or highball, and Paul Harvey at midday. This peacefulness and arms-embracing the world is the king of grist of which world peace is made of.

"This special Mother cultured us in many ferments. She led us to football, dancing, opera at Central City, museums in Denver, and family trips that still warm the heart. Mother Portia gave little nudges and encouragements, but was never demanding. Her generosity was overwhelming.

"While I was never one to court danger, I found my share. Be it BB rifles at school windows and street lights, or mud balls and plums at cars, or riding driftwood in the flooding South Platte, Mother was always there to admonish sternly, occasionally paddle, but never without sustaining Mother love. There were four of us boys and in later years we rather smiled at her idiosyncrasies and even neuroses, happy that she was as well put together as possible, given the stress we boys had flung on her. This beauty, this love always nurtured and always sustained us.

"A difficult task befalls child in giving back to parent. There was no way I could ever repay her gifts of life and days of joy. I think she was proud of us and me! The day I graduated from medical school her face and heart were bright. She loved to visit and share all of life with her granddaughters. Oh, the days of rosy cheeks and hot chocolate at Powder Pandas! Beautiful to recall this and more. The embroidered 'TieHack Kid' ski racer's bib, clothes for Alexi, house rules that mandated clean plates were 'Grandmotherisms' that touched us all.

"My mind is gorged with memories too numerous to recite. The recollections are what immortality is all about. The brain's DNA holds this knowledge and a perfect map for carrying on Portia's beauty, goodness and motherhood. This is how I felt the morning she died. Having dealt very well with a cancer that takes many in less than a year, she made it through birthdays, family reunions, trips, and even skied! I didn't suspect her dying this way, but I sensed some change. Pam and the girls had visited and 'balanced her with Edu-k,' but there was a subtle sleepiness on Sunday, some dozing off and talking of Pops, Luby, and Pete. 'Are you dreaming, Mother?' I asked. 'Unhuh,' she replied. Strumming some quiet guitar tunes, I realized her passing, metamorphosing change was at hand. Nothing changed and midnight passed and the wee morning hours came. Fingers tired, I lay down the guitar and dozed a little. Birds chirping in the early dawn, I felt a slight chill, and heard her take some deep breaths. I started to her, knowing it was time. A trace of dawn coming, birds (she loved them so much) awakening. Yes, this is that peaceful time of day—the hour of the swan. I asked the nurse to get Pam and she came quickly. We embraced and gazed down on this magnificent mother breathing, but a few more times. 'Is this really happening,' I asked myself disbelieving. 'Yes,' I said, 'it is.' My mind raced over all our lifetime. Who will wash raspberries from your face, or tie your hat on, or cook your supper or make you laugh, or always be there to uplift and care? Do you know how very very much I love you? The answers were clear in spirit and permanent, yes, even immortal. 'It's okay mother, you can go now—we are going to be okay,' and she softly passed on. This soul, this woman of many seasons, this mother was gone. I bid peaceful passage to the next life. Her relative body and self she vacated, but she carries on as the absolute, immortal Mother of my life.

"The room was calm now. We sat quietly with her for quite some time. I felt warm and noticed a very soothing orange golden glow. Ah, yes, life and Mother start again this beautiful cycle, never ending.

"Paul C. Curlee M.D."

[The following was submitted by son, Charles K. Curlee, Midland, Texas.]

MY LIFE WITH PORTIA

By Charlie Curlee

"Let's have a picnic at the Slaughter House Bridge!" The enthusiastic cry would burst forth stimulating a flurry of spontaneous activity. "Someone call Cindy. Francis, bring me a cooler. Charlie, help me make some sandwiches." The orders would fly fast and furious and we all jumped at the charges, scurrying through the house, bumping into one another in the narrow passage of the ship's galley kitchen, begging pardons with smiles and excited anticipation. The sandwiches were always simple, perhaps with thinly sliced teriyaki steak left over from the night before, along with a few chips, Frito's or canned peanuts, maybe hard boiled or deviled eggs, pickles, beer. In no time we would be piled into a car for the quick drive down to the Roaring Fork, a short hike along the shaded foot path, soon to be spreading the impromptu lunch on a grassy, sunlit bank along the sparkling, clear, singing river. Such was the essence of "life with Portia" in Aspen: Enjoyment of the simple things in life—the natural environment, warm friendships with new and old friends alike, love of family. With the laughing roar of the Fork for our background music, we would eat, tell stories, wonder at the mountains and hillsides around us, marvel at the lushness of the riparian trees, point excitedly at the glimpse of a Western Tanager or a Belted Kingfisher darting by, laugh at the chatter and antics of the gray squirrels scolding us for coming into their yard. Later, we would slowly stroll along the Rio Grande Railroad grade, continuing our visits, bird watching, and laughter.

On another day the spot might be Herron Park, at the other end of town and equipped with picnic tables if we were going to get fancy and pour wine into plastic glasses. Or it might be a trip up the Pass—to the Grottos or Weller Lake—or up Castle Creek to Ashcroft and fishing after lunch or exploring the old ghost buildings of the mining town. No wonder we all jumped at the charges—the picnics were always delightful explorations in unparalleled scenery! Her glee and excitement were contagious, her energy towards these endeavors seemingly boundless, her smile in that environment as bright as the high country sunshine.

And always game. I first learned of her pluck as a young boy growing up without a father. She would arise before dawn on fall weekend mornings to go sit in a freezing duck blind so all her sons could pursue the sport of hunting. During the early years in Aspen, she continued that outdoor orientation, as if the indoors could not have contained her flair. She took up alpine skiing at age fifty and with arduous determination, mastered the sport so that as a grandmother she would lead packs of parents and kids down her favorite runs at West Buttermilk: Blue Grouse, Camp Bird, and Teaser. The graceful ease of her broad carved parallel turns belied the earlier struggles from snowplows, to stem christies, to finally getting "over the hump." Several Highlands ski instructors could entertain an *apres* ski crowd for hours with tales of those early trials, I'm sure. Later, we would all nod in knowing approval at the cutout cartoon taped on the kitchen fridge in which one youngish, tired skier admits to the other stylish skier who's still going strong, that she's quitting for the day: "This is going to be my last run. Okay, Grandma?" Life with Portia.

Winter was only one seasonal side of Aspen. Summer was her elemental nature and amid her gardens, flowers, birds, and entertaining, she continued her gameness for old and new endeavors, alike. She loved to entertain and would put forth much exertion preparing gourmet dinners, as well as organizing any innocent bystanders into a crack work force to set the tables, polish silver, pick flowers, fold napkins, or perform any of the myriad of other tasks required to host a perfect dinner party. (I believe she could have capably managed a Five Star Restaurant!) And the real fun would commence after supper around the bridge tables with the stories, laughter, and antics from this party animal.

Out of doors, she sweeps around the yard in brightly colored shorts and shirts, a floppy broad-billed straw hat and and infectious smile. Although she could commit to short hikes, she could never have been mistaken for a back country mountaineer. She would leave those treks for others, not her style. On more than one occasion, she stopped at Weller Lake, content to enjoy the day by capturing the glorious moments in watercolor, while others climbed into the land above the trees and to the surrounding snow-covered peaks. At Bunny Court, her outdoor passions were spent recreating Maui as a rock garden in the backyard, along with planting a new tree to honor each new grandchild.

Handles for fireplace tools. The client could not decide which animal head, so we settled for three different ones. Sea horse, forged from 1/2"x1-1/2" stock, owl from 1-1/4" round, elephant from 1-1/8" square. 1975.

Still, one summer I was able to talk her into a day fishing trip up Independence Pass. I equipped her with a set of waders and off we went to a spot where I was certain she'd be able to catch a brookie on a fly. My early memories of the Roaring Fork Valley include my mother taking the day by pulling in several fat rainbows on salmon eggs from one big hole by a large pine tree next to the parking area, much to the disgust of the fly fishermen in the group and a story she always recounted with chuckles. She had spoken occasionally of wanting to learn to fly fish and this was a game opportunity.

From a highway pull-off, some distance up the pass, we proceeded to slip and slide down slope through the Aspen grove, stopping every once in awhile to admire the asters, columbine, lupine, Queen Ann's lace and other wildflowers, eventually dropping onto the valley floor far below the traffic on Highway 82. We found a secluded opening in the stream, prepared the leader with a dry fly, and waded in below a quiet, elongated pool, where occasional circular ripples gave us the telltale sign of feeding trout. I let her work the rod, knowing that she had done some fly casting, but that it would take her a little time to get the hang of it again. Miscasts had me fetching the fly from a scraggly spruce along the bank behind us, from around a soggy branch along the edge of the stream, whoops!, replacing the fly that just got popped off on an errant backcast. Eventually her casts leveled out, straight into the pool, but without much distance. (Watch the fly now, Mother, you'll need to set it just when you see the trout suck it in. There!) But too late. Another cast, another strike, another miss. Several similar repeats with breaks to untangle snags, tree, trout, other misadventures. Then the hole was still. Spooked brookies. We worked upstream slowly, fishing some, visiting mostly, enjoying one another's company, reminiscing, laughing. As we crossed the stream to get a better angle at the next pool, the inevitable happened. The rocks on the bottoms of the mountain streams are generally slippery, even to the felt soles of our waders. Add to that the strong current, shifting rocks, and an older woman whose innate agility has waned, and Kerplop! With a surprised, high pitched scream, she sat down, almost gracefully, in six inches of icy Roaring Fork water!

The game lady was all right, only bruised pride and a cold behind. A little flustered but still full of laughter, smiles, and pluck. We continued to fish a while longer that day, but she never managed to catch a brook trout on her own casting, much to our mutual disappointment. Towards dusk we trudged up the steep slope, stopping several times to catch our breath, happily contented at having experienced together a day of glittering sunlight dancing on the stream, of bird song and mountains, aspens and spruce and birch and wonder, sharing a small slice of life together in an idyllic setting . . .

As years went by the game lady moved her passions inside more and more. Exactly when she began serious needle work I can only guess, but the same determination that helped her master the motor control of skiing, helped her become an exquisitely skilled quilter. With every visit to Aspen we were shown any of several projects carefully laid out in various stages of incompletion: patches of multi-colored cloth cut out in what to my untrained eyes were random scraps, an assortment of carefully folded patterns, several pieces of cloth fragilely held together with basting stitches, ribbons, thread, needles, scissors, pins. We would sit peacefully at the breakfast table relaxing over a last cup of morning coffee, enjoying the show of birds at the backyard feeding stations, when she would pull out yet another project from a bag at her feet, gently unfold layers of cloth and matting, and begin precise, detailed stitching. We marveled at how she could ever keep them all straight. A vest for Francis, the cathedral window wall "cats" for the master bedroom, a king-sized quilt, vests for each grandchild, a new sewing bag, ski bibs for the kids, and on and on. And we marveled that she completed any of them amid all of her other daily chores, activities, commitments: processing scraps for the compost bins, triple spin-washing garden lettuce, making jams and jellies (wild plum, fresh picked), putting up chard for winter consumption, visiting with neighborhood friends, and traveling with Francis to blacksmithing workshops, CU football games, exotic vacations—Whew! But the prolific projects were completed and others took their places in the baskets, sewing bags, and tables around the house so the next visit to Aspen would start with a tour of all her current projects.

Then miraculously at Christmas time, there for family and friends were carefully sewn tree skirts, Christmas potholders, gloriously quilted Christmas stars, bright red and green wreaths, and lovingly (and secretly) cut, pinned, stitched, tied, wrapped and mailed out along with the traditional sixty proof brandy socked fruitcake. Gifts crafted with love through all the seasons of the year, for the season of love, to be enjoyed throughout the years. This was life with Portia: simple loving enjoyment of family, friends, life.

[Submitted by son, Charles K. Curlee (Charlie), Midland, Texas as told to Francis Whitaker.]

Late in 1975, Portia became unusually anxious about the daily mail. The postman came around noon, when I would be coming from The Mountain Forge. Often she would wait to get the mail, if it came early, or, if it was not in when I got home, would say, "Don't bother with the mail when you go out, I'll get it for you when you get home for happy hour."

This went on for some months, and then ceased suddenly. All was quiet on the postal front. Then, one day she said, "Helmut and Brunnhilde are having a fancy party next Friday, so please wear your white formal jacket and black trousers. I was puzzled, but I reluctantly agreed. Friday came and when we got to the Schloffer's house the place was filled with many of our good friends in Aspen, plus a photographer. (I hoped this was not another of Portia's practical jokes.)

At the appropriate moment Portia called for silence and proudly announced that I had been awarded a Doctorate in Humane Letters from the University of Colorado! After living with Portia for over twenty years I, of course, thought it another of her elaborate practical jokes. When the tears started to come into her eyes, however, I realized that what she said was true.

All these months, she had been keeping her efforts a secret, while she conspired with her favorite professor, as well as many of my long time friends and customers, to get letters of recommendation from all of them. At first, I thought she was joking, not unusual for her, but I finally realized what she had been doing. She had been keeping her huge secret from me and swearing secrecy from all those concerned.

It worked. In the spring of 1976, during the graduation ceremony on the University Campus in Boulder, I was awarded a Doctorate in Humane Letters. The high school drop-out had become Doctor Whitaker! This, of course, was with thanks to a most wonderful woman. Portia also presented me with a scrap book containing all the secret correspondence and was completed with pictures of the ceremony and me in my cap and gown. I couldn't have been happier. All the years of hard work and dedication had paid off again.

7

MEMORIES OF MY FATHER IN THE SUMMER OF 1942

By Stephen Whitaker
April 1993
Davis, California

It must have been the summer of 1941 that Jim Harkins signed on as a cook for the Sierra Club High Trip, because his tales of the High Sierra caught the fancy of my father and we went to see for ourselves in the summer of '42.

I suspect that the two of them began making plans for a pack trip many months in advance, since equipment was scarce in those days and it took more than a quick trip to the local mountain shop in order to get outfitted for a serious hike.

A couple of years earlier my parents had made sleeping bags for the entire family so that we could be comfortable at the Sierra Club Lodge near Donner Summit during winter ski trips, and these wonderfully light goose down bags were perfect for summer pack trips. The other equipment, such as pack frames, tents (we had one small one for emergencies only), stoves, and dried food were scavenged from Army surplus and local markets. Jim's experience as a cook on the High Trip wasn't much help with the food, since the Sierra Club had used mules to transport a rather luxurious commissary over the passes and along the trails. In our case, we would carry it all since my father was partial to being independent and the idea of traveling with a horde was not to his liking.

In later years, as the number of High trips grew, he would wait until the Sierra Club published its schedule of trips for the summer in order to determine where *not to go*. However, in the summer of '42 the mountains were virtually empty, and it was unnecessary to play hide and seek with the High Trip crowd. Here I should point out that my father was an ardent supporter of the Sierra Club and has been a lifetime member for more than half a century. With the Sierra Club he became involved in many conservation projects which are too numerous to list here, but when the time came for the summer wilderness experience, he didn't want to share it with the crowd and the meals by mules. The winter wilderness experience was treated in much the same manner and in order to haul heavy loads on skis he invented a combined cross-country-downhill-safety binding that took us far from the crowd. But that's another story.

The year 1942 found us moving from Carmel to Mill Valley so that my father could work as a welding instructor in the shipyards, first in Sausalito and then in Richmond. Paid vacations were unknown in those days, and in order to take a vacation without losing your salary, you had to put in the extra time before you left. The standard work week was Monday through Saturday, and we saw very little of my father for two months as he accumulated eight days worth of "paid" vacation. The trip from Mill Valley to Richmond required the use of a ferry, thus my father left the house before the rest of the family was out of bed and he would return late in the evening to eat dinner and continue preparations for our trip. Preparing food was a difficult task, since few people knew how to travel light in the mountains, and food in general was in short supply because of the war. My contribution would be fish, and I was eager to try out a newly acquired fly rod and a wonderful assortment of dry flies. Earlier my father had taken me fishing in the small streams along the coast south of Carmel, and, of course, the Carmel River. A pole, a line, a hook, and a worm or salmon egg were all that one needed, and trout were available in the tiniest imaginable streams in the area around Carmel. When I return to these streams today, they are nothing more than a trickle, but in the old days the trout were there and could be taken by those who wanted to work at it. A fly rod and artificial lures was another matter altogether, and the idea of big streams and big fish provided plenty of food for a fertile imagination.

In August of 1942 we left Mill Valley early one morning on our way to Florence Lake and the trip to Evolution Valley. Since we were coastal people, accustomed to cool summers and mild winters, we feared the heat of the San Joaquin Valley and the crossing was always made early in the morning. Air conditioned cars were nonexistent in those days and one survived the heat by frequent stops at the ubiquitous Giant Oranges that strung out across the valley. There one could have fresh squeezed orange juice poured over ice to soften the pounding rays of the sun; however, the Giant Orange was an expensive proposition and during all our trips across the valley, I remember stopping only once. In our family you didn't pay a buck to beat the heat, you got up early and hit the road!

Florence Lake was actually a PG&E reservoir and we arrived there in time for a picnic lunch, after, of course, we had strapped on our loads and followed the trail around the lake. As the youngest hiker (I was ten years old) my job was to carry the four very light down sleeping bags along with my own clothes and fishing equipment. By any standard my pack was light, but I didn't like it one bit and within half an hour I was audibly suffering. On the lake there were several motor boats speeding around with ease while their occupants enjoyed the intense sun of the mountains simultaneously with the cool breezes of the lake. The trail was a different matter. We had indeed escaped the valley heat by our early morning departure, but the trail around Florence Lake was hot and dry. We were hiking several hundred yards above the lake shore. Trapped between the dusty trails at my feet and visions of the cool breezes being enjoyed by the boaters on the lake, I was, quite frankly, miserable. Eventually, we worked our way around the lake and headed deeper into the mountains where we would see no one, except a single person of the trail crew, for the next eight days. My misery on that dusty trail during that hot afternoon would later turn into one of the warmest memories of my childhood. These memories would take me back to the High Sierras regularly during the next half-century. Curiously, I would never return to Florence Lake or ever again set foot in Evolution Valley, yet the memory remains.

Our first campsite was nondescript, that is to say that I remember nothing save that it was located along the San Joaquin River in what would be described as the lower reaches of Goddard Canyon. The altitude was relatively low and the trees were large so one did not see much of the surrounding country. Packs were very heavy during the first part of the trip and my father, Jim Harkins, and George Smith engaged in the lost art of "shuttle packing." With the equipment that is available today, one can virtually float through the mountains like a butterfly, but in 1942 everything was heavy and the loads were more than the group could carry at one time. So my father, Jim, and George would load up and head off while my mother, my sister, and I would follow slowly. The men would deposit their loads along the trail a mile or two ahead and then return to the original site and pick up what had been left. After the first three days we had consumed enough food and had become accustomed to the lighter loads (in my case the weight never changed and I let everyone know that this was grossly unfair) so that shuttle packing was no longer necessary.

While the original loads were indeed excessive, they would have been intolerable if my father had not been one of the early masters at *going light*. For cooking he had made two utensils that took care of virtually all our needs. From one of the local restaurants, he had picked up a couple of two-quart cans which had originally served as containers for fruit or jam. To these he added a wire handle to make a pair of "billie cans" that served us well for many, many trips. These were used for boiling water, making soup, and cooking oatmeal. The outside was allowed to blacken over the wood fires, and because of this they had to be carried outside of one's pack where they were constantly banging into each other, in addition to every rock and bush within reach. They were truly a nuisance to transport, but they were light, cheap, and effective. The second utensil that my father made was a marvelously sturdy aluminum grill. This had slightly turned up edges for retaining a bit of grease when pancakes or trout were being cooked, and it had been fashioned from one quarter inch thick aluminum plate in order to provide a uniform temperature when heated over a wood fire. We carried the billie cans and the grill, along with a complement of Sierra Club cups and a few spoons and forks, and that took care of all our cooking needs . . . in terms of things that had to be carried. To this day I have never seen a commercially available grill that is the equal to the one that my father made for the mountains of a different era. Because of the increased population in the mountains, I have not built a wood fire in nearly two decades and most thoughtful hikers have accepted this change in wilderness camping.

It should not be surprising that my father was a master of the cooking fire, since the forge fire is as essential to the blacksmith as the hammer and the anvil.

His technique consisted of selecting a vertical rock surface, and then building a two-part kitchen fire as I have illustrated in the drawing below.

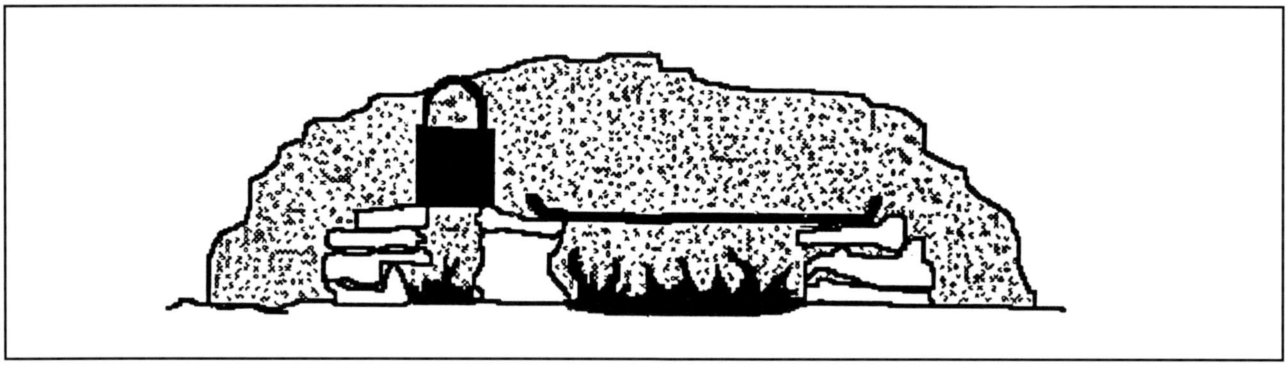

Francis Whitaker's High Sierra Kitchen

The small fireplace on the left was for heating water or making soup, while the larger space on the right was for pancakes in the morning and fish in the evening. The space on the left was always built to accommodate two billie cans, so soup could be prepared in one can while hot water for noodles was heating in the other. My contribution to this activity was to gather wood, and this was a bit of a chore since a selection of three sizes was required. To begin with, one needed very small, dry twigs that provided a small fire which would be used as the starting point for the main firewood, which was never more than an inch in diameter. Once the fire got started, the rocks held in the heat and only a small amount of wood was required to keep the kitchen going full blast. Given the number of people that are using the mountains at this time, the Whitaker kitchen is no longer acceptable. In the early years, however, it was the most efficient arrangement that I and others had ever seen. My father was clearly the master of the High Sierra, wood-fired kitchen stove.

Our second day of our trip along the San Joaquin was one that I shall never forget. As we slowly gained altitude, the country opened up and by the time we reached the falls that spilled out of Evolution Valley the river was no longer closed in by trees and brush and there was enough space for some easy fly casting. About casting a dry fly, I knew nothing, but George Smith had some experience and as lunch was being set up, he began fishing in a big pool next to our resting spot. He was using a wet fly and had picked up a couple of strikes when he offered the rod to me. After a few practice casts, I was able to put the fly where George had suggested and as it drifted with the current, he told me to be ready for a strike. Bang! . . . the fish hit just like he had said it would and I reacted with fury . . . pulling the pole up over my head with such force that I broke the tip of George's fly rod. By the time I got untangled, the fish had worked its way loose and I was very disappointed. Never before had I felt such a big fish on my line. When I had the chance, I had blown it.

George was not happy about the broken tip. Fortunately, I had a spare tip that had come with my newly purchased rod, so we could both continue fishing during the trip.

I have always wanted to go back and try that pool one more time, but it has never happened and I have had to live with only the memory of the "big one that got away."

After lunch we climbed the falls out of Goddard Canyon and into the lower reaches of Evolution Valley where we found the first of what would be many, many idyllic campsites. This was our first truly classic High Sierra campsite . . . located right next to the stream with a grand view of Evolution Valley toward the east and the lower reaches of Goddard Canyon toward the west. My young life was filled with such campsites since my father had high standards for his mountain accommodations, and he was always looking for a "room with a view" and, if possible, one with a breeze that would keep the mosquitoes at bay. Today these campsites are off limits, since one must camp at least 100 feet from any stream or lake. I cannot argue with this necessary policy, but I am grateful that I once had the pleasure of a bedsite with an "up close and personal" view of the local stream.

At this point in our trip, I began fishing in earnest in the seemingly endless series of pools that stretched out across Evolution Valley. For the next two days I fished, my mother and sister rested, and the men scouted the territory ahead and prepared to set up a camp at Colby Meadows, our final destination. The trip to the upper

Second Yellin Workshop. Detail of gate crest. Heads forged from 2" square. 1986.

reaches of Evolution Valley and Colby Meadows was fairly easy in terms of hiking, but rather agonizing in terms of the battle against the mosquitoes who came in swarms to feast on the most *penetrable* flesh they could find. No question about it . . . that flesh was mine. In 1942 the plethora of chemical repellents that we know today was non-existent, and the mosquitoes considered the lotion on my skin to be something of an hors d'oeuvre. On more than one occasion, frustration got the better of me and I swore that I would never set foot in the mountains again. My father was not seriously bothered by the pesky mosquitoes, since they found his tough skin much less palatable than mine, and I have lived through this same cycle with my own children as the clouds of mosquitoes in the 70s followed their ancestor's instincts and dined on the soft bodies of my children rather than that of their father.

In Colby Meadow we camped by a small feeder stream so that we would be above the main meadow and its army of mosquitoes. A little altitude in that type of situation means less moisture, more breezes, and fewer mosquitoes. years after I watched my father search for that perfect combination of water, wind, and view, and in my own time I found some rather enchanting campsites to share with my family by following his persevering style.

Sometimes water, wind, and a view were not enough and he would continue his search until he found a site that had a bathtub nearby. During our early years in the mountains we took the plunge into many an icy stream, and today I can only wonder how we withstood the cold.

The fishing in Colby Meadow was truly fantastic and I provided an endless supply of fish for dinner after dinner. At first this was a welcomed treat, but after a day or two the rules of the game changed. Originally everyone had helped prepare the daily catch, but then the rule of, "you catch it, you clean it" was imposed and I began to be a more selective fisherman. While I spent the days fishing, the men spent their time contemplating a climb of Mt. Darwin which, at 14,000 provided a significant challenge. I was adamant that I could make the climb, and arguments ensued; however, on the morning of the climb the men slipped out of camp without waking anyone and I was left to fish in the solitude of Colby Meadow. I was quite furious that I had missed the climb, and vowed that someday I would do it myself; however, I have never made it to the summit of Mt. Darwin and I suppose that I never will. It remains intact as part of that original pack trip . . . never to be revisited, always to be remembered.

There is a saying that it never rains at night in the Sierra. Some people are more flexible and say that it never rains all night in Sierra, but in Colby Meadow it did indeed rain one night. In 1942, a tent was something made of heavy canvas that you carried around in a car, or perhaps a truck, and the modern lightweight tent was unknown. Our group of six carried one small, heavy tent, and when the rain came we had to make do with whatever was available. I don't remember how everyone else survived, but my father found me a bit of a cave and I spent a rather comfortable night. Given the equipment that is currently available, I almost always carry a tent but aside from that first trip in '42 I can't remember going on a trip with my father and carrying a tent. A ground cloth

A study in ovals. Grille made at the first Mountain Smiths Conference at The Colorado Rocky Mountain School, Carbondale, Colorado. Frame of 1/4"x1", ovals of 1/2" square, scrolls 1/4"x1/2". 1990.

or a cave would provide the necessary protection, and if that failed, you could simply remind yourself that it never rains at night in the Sierra.

Eventually, food and time ran out and we had to descend back down Evolution Valley and Goddard Canyon. What had taken us four days to achieve, we would undo in an easy two days; however, the second day of our descent was not easy for me. As we worked our way down along the San Joaquin River, the hike became progressively more difficult for me and eventually I had to give my pack to my father. I was feeling quite sick, and I wondered if I was going to make it back to Florence Lake without being carried. My memory of our exit is not very clear, but I have visions of myself saying, "Never again, I will never do this again."

Back in Mill Valley the diagnosis was measles and I spent a week recovering in a darkened bedroom. With time, the image of an armada of mosquitoes feasting on a helpless boy disappeared, and memories of the wonderful solitude sharpened. A year is a long time in the life of a ten-year-old, and in the summer of '43 we were off to Lake Ediza for more of the grandeur that was the High Sierra of my boyhood. But that is another story.

—STEPHEN WHITAKER

8

MY TRIP TO EUROPE—1992

Next time, take the train, especially a Swiss train, but more of that later. When I got to the Grand Junction airport on June 1st, to take the first of four flights, Denver, Dulles, DeGaulle, Bordeaux, the first flight was posted "Delayed." With only half an hour in Denver, I was concerned that if I missed the second flight, I would be in trouble all the way. I tried to get United Express to put me on Continental, that jet flight, ready to go on the runway, which would put me in Denver within forty minutes. "Oh, the weather in Denver is heavy, all flights are delayed." It was a bright clear morning in Grand Junction. All flights but mine, which left on time—I missed it by five minutes!

The next flight was via Chicago, but I'd miss the 9:00 a.m. flight to Bordeaux. My bags?? Where and when would they be? They were checked to Paris, a mistake, as they should have been checked to Bordeaux and go through customs there instead of Paris. No way to reach my son, Stephen, and his wife, Su, their phone number was in one of the bags. Got to Paris on time, but the Bordeaux flight was from a different terminal, involving a bus trip and a long walk, an hour minimum to connect, what with claiming my bag and going through customs. Only one bag made it to Paris, the other one went to Los Angeles. I got four hours sleep from Chicago, but made the 12:15 flight to Bordeaux. Stephen and Su were waiting, the French airline told them that I was on a later flight. Good for them.

A charming hotel in Bordeaux, room on an inner courtyard, overlooking a trimmed olive tree with a trunk three feet in diameter. Got a three hour nap. A fine dinner, blackened sole for me and a large bottle of local wine. Off tomorrow, IF my second bag arrives.

June 3rd. A 9:00 a.m. call from Etienne, as Stephen is called here, revealed that the lost bag had been sent to Paris and would arrive on the 1:25 p.m. flight, a full day late. It would delay our departure for St. Etienne de Baigorry, in the Basque country, but as it is only a four hour drive, we should be on schedule from then on. A lazy "petit dejeuner," watching Bordeaux pass by, especially interesting to see the people with unwrapped three foot long loaves of bread under their arms. A note on the French telephones. Stephen told me that even if I had his number, I would have great difficulty in phoning him. All French telephones are electronically, not coin-operated. A card is inserted in the phone, and its value is automatically reduced as it is used. Such cards can only be purchased at specific locations, and I had no French money! Hallelujah, my second bag arrived.

So, we were off at 2:30 p.m. Out of Bordeaux and onto a four lane highway headed south. Miles and miles of planted pine forests on either side reminded me of New Zealand and their planted Monterey pines. Halfway to Bayonne, we turned off onto local roads, and the flat plain slowly turned into rolling hills, lovely winding roads through small towns (now with closed shutters, and the corner stones on the wall exposed). Light rain off and on, green hills growing bigger, rushing rivers, swollen and muddy from recent rains, we traveled on. We did get lost once.

On to Hotel Arce, in St. Etienne. Charming place, right on the river, my room had twelve foot ceilings, papered and decorated in rose. Elegant dinner, pension style, with champagne first just to celebrate our being their. Soup, salmon, pork tenderloin, salad, and dessert. Two hours for dinner, a twilight walk along the river, and then to bed.

June 4th. A side trip to St. Jean Pied-de-Port. The name means at the foot of the gate or entry into Spain. Left the car outside of the Ville Centre, a practice we continued throughout the whole trip. We walked up to the fort on the hill. It was constructed of amazing stone work, ports for bow and arrow and other weapons, but the now peaceful and beautiful gardens were all that was left of a very warlike period. House lots were about 40 feet wide by 300 feet long, festooned with flowers, vegetables, and fruit trees. We did a bit of shopping for Basque linens. Visited a Basque church, had lunch in Vambo, covered a lot of ground on foot with my camera and returned to our hotel. It was still drizzling. A two hour nap, as I was still catching up in my jet lag. Spain tomorrow, over the Pyrenees.

June 5th. Clearing this morning, the rain has stopped. Off at 9:00 a.m. A steady climb on a steep winding road. We showed our passports on the French side, but no Spanish checkpoint. Su told us that the borders were now wide open. The country changed quickly as we descended to Pamplona. The lee side is gentler and drier. Houses not as neat, some need whitewash, and the land is not as carefully tended. Reached Pamplona by noon. The central plaza jammed with cars, double and triple parked. Stephen found a place to double park and went to a bank to get Spanish pesetas. We had a hard time getting out, no left turns for a mile. Got on the freeway, they are all toll roads, headed the wrong way, but quickly got on a red road headed for Olite. The roads are all color coded on the highway signs and the road maps. Blue for toll roads, red for national roads, yellow for secondary roads, and white for country. Very simple and effective. We found the correct road and were soon at Olite. A magnificent castle awaited us, part of it converted to a Parador, one of the state operated hotels. We went looking for lunch and a man on a scooter guided us through the narrow, walled streets. We managed to get a light lunch. There was not much between a simple bar with snacks and an elegant restaurant in the town. After lunch, I covered the old walled town with my camera. There were lots of balconies and door grilles, from the old days and some of them were good cast iron, however the arc welder had already reached there and a lot of it was cut and paste.

June 6th. Breakfast reminded me of the Bergen Line last summer, the whole works, not the petit dejeuner of France. Eggs, cheese, ham and other cold cuts, fruit juices, croissants and breads, cold cereals, and stewed fruits. I soon learned that there was little ironwork in the 11th to 14th century churches and monasteries. Marvelous stone work, vaulted ceilings, grotesque figure heads, but all of a dull brown stone. The whole town is stone, the streets so narrow we had to fold in our side mirrors.

Most of the land is in grain or grapes, the grain fields already golden, the grape vines cut very low, although the trunks showed old age. Many hillsides showed ancient terracing, often with planted pine trees. Twice, we saw men with sticks breaking off the long grape tendrils, so as to keep the growth concentrated. We are still two to three thousand feet above sea level, so it is warm in the sun, but cool in the shade or inside the stone buildings. Tomorrow, a long day, 200 miles to Segovia.

June 8th. Not a long day after all, we were in Segovia by two. Spanish roads are well marked, the directional signs extremely clear and the color coding system excellent. A steady climb from the golden grain country. Cold and cloudy. We stopped at one small town and found a charming secluded square, a small church, with some of the heavy solid Spanish ironwork. This is stork country. They have their huge nests on top of all the towers and church spires. We had a tricky time finding a hotel. It was built into a steep hillside, the garage was on the bottom level, with the entry and reception on the 5th level and our rooms were on the seventh level. I had a sweeping view of the Spanish countryside, a monastery in the distance.

What ironwork! The Cathedral Plaza had over 100 balconies all of excellent traditional ironwork. At every turn on the narrow cobblestone street there were more and more. It was risky walking. You hear a car or motorcycle, the sound echoing like thunder in the mountains, so you squeeze against the wall until the vehicle passes, or it goes another way. A sunny morning, with a light thin overcast, much better light than yesterday, and soon I realized that I hadn't brought enough film. The light and shadows were perfect for showing off the old ironwork.

Segovia, a city of stone and iron. Stone buildings, stone streets and iron balconies, literally by the thousands. Profuse railings, gates and wooden doors were all over the town. Only the doors could not be made of stone. Intricate knockers, handles, and many variations on decorative nail heads and bosses graced the old wooden doors. The Roman Aqueduct was a marvel and still is! Huge stones, apparently fitted without mortar, centuries old and still functioning. How were these huge stones, weighing half a ton and more, put into place? A close look showed a small indentation on every stone, at or near the balance point. There must have been huge tongs to grip the stone by these indentations, with scaffolds, ropes, and unlimited manpower. To further prove my theory, the tapered stones forming the arches, had indentations placed so the stone would hang correctly to fit the arch.

What an example of workmanship for generations past and present!

June 9th. To Avila, a slow climb up another 1500 feet. Cold and cloudy and 48 degrees. Snow was falling in the Pyrenees, so we were fortunate with weather when we had come through. A different style of ironwork became apparent in Avila. Crests were now on the window grilles, and new patterns on the balconies. Unfortunately, no photographs were allowed in the churches or castles, so I could not record the marvelous ironwork. I got lost a bit returning to the hotel, and found a blacksmith shop. Alas the forge was dust covered, the arc welder, the grinder, and the Hossfeld bender had taken over. It was what I expected after seeing some of the contemporary work. Very sad, with infinite inspirations everywhere, to see a great art lost. It makes me all the more determined to preserve the traditional techniques. A lost cause? *Quien Sabe?*

June 10th. Off to Toledo. A gray morning, and our route was over a mountain range, the top of which was in clouds. Excellent road. A climb into the pine forests, with fog, dense at times, it might have been on the Monterey Peninsula. The pines changed shape, branching at medium height into umbrella like tops. As we crossed the pass, the fog cleared and the sun was brilliant. A long drop down to El Excorial, the site of a famous monastery. We spent a short time here as there were hordes of people lined up to enter. The town had little charm, so we were soon on our way to Toledo. At noon, we found a tappas bar in a small town, parked right in front of it so we could watch the car. A sandwich and a beer, then I walked over to a small church. I found a big reja, and since men were working there I tried a flash shot—it worked. Traffic was light until we hit Toledo. A minor accident had traffic snarled so we made a slight detour. We could see the hotel, but we couldn't get to it because of the traffic backed up. Stephen parked a few blocks away, and walked with Su back to the hotel and got directions. By the time they returned the traffic had cleared, so we drove to the hotel.

Part of the wall around the old town formed one side of the garden and a huge mansion, the Hostel del Cardenales, the residence of a former Cardinal. There was a lovely garden with trees, flowers, and a fountain. My room was very quiet, it looked out onto an inner courtyard. I went exploring alone through the narrow streets. I saw some excellent ironwork at one of the churches. How different our various lodgings had been. The weather was now very pleasant as we were farther down the mountains as well as being further south. Perhaps the unusual cold spell had ended. Swallows filled the air in the morning and it was a miracle that there were no collisions as they flitted and dove around each other. I am trying to be more selective about my photographing, but I have already made over 200 exposures!

June 11th. Toledo! Shirt sleeve weather. After the usual continental breakfast, off to explore. When we reached a hotel, we almost never used the car until we left. We walked everywhere. First we stopped at a bank, but Stephen had left his passport at the hotel, and the bank would not cash his travellers checks without it. So we went back to the hotel and found that he had left it at the desk. This was a common practice in Spain when one checked in. Up and down the narrow streets we walked heading for the Cathedral. Traffic was horrendous. I often had to duck into an entry way for safety! Anywhere there is half a space to park a car–a whole car was parked there. How lovely this part of the world must have been before the automobile. My father had told me of his trip here sixty years before and how wonderful it had been. He told me that they used the train, and often walked from one small town to the next. One of my great memories is his telling me of walking from France to Spain.

The Toledo Cathedral was magnificent, and as seems to be the custom, the fine ironwork is almost unrecognized. In all of the guide books, there is only a brief line about a smith who made one of the rejas or grilles and did not charge enough. Here is a quote from a Toledo guide book, "Master Domingo de Cespedes, a Toledian grille maker, who in association with his brother-in-law, Herman Bravo, created this collosal enterprise (1541–1548).

Replica of a Yellin Grille. About 7"x10". Grapes forged in a die, two at a time, all forge welded together. 1985.

Master Domingo undertook to make the grille for 6000 ducats, but the job drew out and he had enormous difficulties in fulfilling the commission. He had not only to sell his properties, but to borrow heavily. His anguished petitions for the grille to be evaluated went unheeded. 'Send to see how I have spent in the said grille all that I earned in my youth, and the large sum for which I am in debt,' he wrote Cardinal Sileco in 1559! The great artist died, completely ruined, an example of honesty and integrity." A 400 year old example of how to charge!! A lesson for all. The main grille as well as many of the memorial grilles or rejas, were masterpieces, yet no postcards, photos, or slides were available, and with photography prohibited, I can only remember their glory. It seems that present-day smiths spend their time making cheap swords and halberds for the tourists. Toledo blades have been famous for over a thousand years, and it is sad to see the craft so demeaned.

Lunch at the Restaurante Placido, the quiet place, and it was. A beer and fried artichokes with ham, bread, and wine for me, and I left Stephen and Su. I allowed an hour to get back to the hotel, but the streets were so narrow and winding, it was hard to maintain a certain direction. It was a pleasant hour, though, some fine ironwork on the way made the trip worthwhile. I went for supper to a local bar. No fancy dinner, they didn't start serving dinner until nine o'clock at the earliest. I had meat balls and French fries, and the fries were marvelous. Spanish tortillas, an egg and potato dish for Stephen and Su, and we found ourselves watching a bullfight on Spanish TV, fortunately, without the sound. The dress and ballet-like technique were fascinating. A nearby church was open on the way back, so we stopped for a few quiet moments. What a strong tradition. The church contained only the older people. It reminded me of Hawaii, there the older ones cherished their traditions, the old crafts, the songs, and the dances. The younger generations went for rock and roll. I asked myself if the old crafts and traditions were fated to go the same way in our country? I certainly hoped not!

June 17th. We had a difficult time getting out of Toledo. For once, the highway signs failed us, but we finally got away. Rolling country, grain, olives and grapes, then into the mountains, crossing the Sierra de Gredos. The high pass was 6000 feet, lovely country, a bit of rain occasionally, snowy mountains in the distance. Many varieties of wild flowers and the hills were covered with a bush glowing with bright yellow flowers. The most colorful and scenic drive yet. In twenty-five miles we met only four cars, a part of Spain that few travellers see. We passed a large National Reserve, and then dropped down four hundred feet on a long winding road to Piedrahita. From there on it was easy driving.

We stopped at Alba and I found some very good iron gates at a Basilica. Su guided us very well to our hotel in Salamanca, in fact her guidance was superb all the way. Our Four Star hotel was a gorgeous place. The rooms had twelve foot ceilings, classic plaster ceiling mouldings and the bathroom was ALL marble. The area was very quiet and only six blocks from the Plaza Mayor.

June 13th. We arrived in Salamanca. How different it was from Toledo. Salamanca is a university town far from

Madrid. It was crowded with young people, no tour busses, and fewer cars. The walking was much more pleasant with not much honking and fumes. The Plaza Mayor, a huge central square, was filled with scaffolding, which made photography impossible, but I did get some good pictures of the myriad of balconies. There was some big celebration coming up which was the reason for the scaffolding and decorations. We did get into one church just as it opened. There was no ticket to buy and no sign prohibiting photographing, so I tried some flash shots of the rejas or railings. We were all a little bit footsore from so much walking, so 11:00 a.m. coffee was in order, and I had a beer. We have solved the eating problem. Since most restaurants do not open for dinner until 9:00 p.m., we have a Plato Combinado around 12:30, salad, meat, bread, and wine, then a snack at a Tappas bar in the evening. Adequate, and I hope to hold my regular weight. This plan was also easier on the budget.

After a leisurely lunch, I went off by myself to photograph the iron grilles on the Shell House. The house is a landmark and one of the best examples of Spanish traditional architecture and ironwork in the whole country. I had to wait an hour for the sun to fall on the wall that I wanted to photograph. The timing of the photography was critical on these narrow streets as the lighting was limited to a few hours of the day.

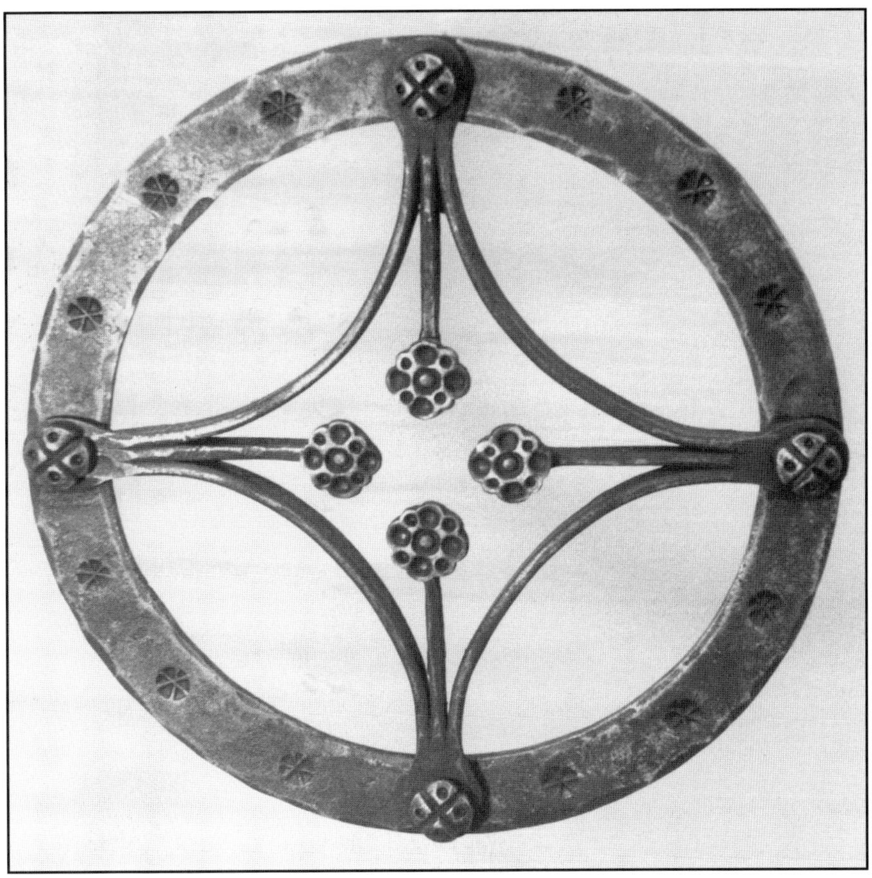

Ring for Bill Gichner's ring project. 1990.

Last night we were startled awake with fireworks! There were high school bands marching through the streets. One group of six musicians, drum, cymbal, and various horns went from bar to bar, serenading outside, playing inside. Su thought they got drinks and snacks each time. Also an old white-haired man, slowly walking the streets, was beating on a drum and playing an odd sort of flute. He was not asking for contributions or anything, he was just having fun. So much for sleeping!

June 14th. We set off for Burgos. We had an easy exit from Salamanca thanks to our courteous hotel clerk, however getting to Burgos was an entirely different story. There is always a big gap between the highway map and the Michelin map of the inner city. Each Spanish town has its own character. Burgos is dominated by the Cathedral. In one plaza we found the possible source for a book on the rejas. We had to order one from Madrid, but at least I have the title and maybe I can order it when I get home. Looking at the old ironwork, I often thought it must have been a wonderful time to have been a blacksmith. The Cathedral surprised us by being gray. All the others we had seen had been all yellows and browns. It was beautiful though.

June 15th. Enter at Olite, depart at Burgos. Such a wonderful way to go. Unfortunately, it was a gray day in Burgos. After a fine petit dejeuner, I was off to the Cathedral. We agreed to meet at 1:00 p.m. for lunch. It took me two and a half hours to get back to the Cathedral. I kept running out of film and having to go back to the hotel to get more. (So much to see and so little time—the story of my life!) I met Stephen and Su in front of the Cathedral and we toured it together. The interior was indescribable. Chapel after Chapel and the rejas were thirty-five feet wide and forty feet high! There were vertical bars of 2" square that were grooved and twisted into every conceivable shape and columns up to 5" round carved with all sorts of allegorical figures. We finally found a gift shop where some postcards were available and we bought two with pictures of the iron rejas.

For lunch we had a delicious charcoal broiled leg of lamb, for which Burgos is famous, a large roll, a gin and tonic, and a media of the white wine of the region. Tomorrow to Hendaye, France, on our way back to Bordeaux. Truly, the grand tour, long enough and of such variety in cities and accommodations. From the luxury of the Hostel del Cardenales to the tiny room at the Hotel Cordon in Burgos. My room was ten feet square, the bathroom six feet square, and I had a two foot square shower in one corner.

I should mention the stature of the Spaniards, from 4'6" to 4'9". The later generations are 5'6" to 5'9". We felt like giants in the crowds with most of the heads about even with our shoulders. They reminded me of the Miyamoto family that we had known in Carmel, California. The Japanese born parents, who had a truck garden just past the Mission, were very short, but Ky and Gordo, their children were husky six-footers. It must be the air or the food!

A final evening. We walked up to the restaurant Su had chosen, but it was not to our liking, so we went back to the big square next to the hotel. This place had three prices for the same food! The bar was the cheapest. If you sat at an inside table the price was higher, and if you sat on the terrace the prices were higher yet—the food stayed the same, only the price went up. It was the scenery I guess. It was hot and smoky inside, Spaniards are awful smokers, so we chose the terrace and watched a soccer game on the street. The players were eight to ten year olds and they were having a great time emulating their adult counterparts—arguments, fights, and all. They take their soccer seriously in Spain. Occasionally the ball would hit a passerby or roll into a restaurant, but the people all took it good naturedly.

June 16th. We awoke to the sound of thunder, followed by a heavy rain. It slowed our trip to Hendaye, 200 miles, but it was four lane all the way so it wasn't much of a problem. Finding our hotel, however, was a problem and took some extra time. We were now on the popular French coast and the traffic was much heavier than we had been used to. The weather didn't clear much on the way and we hurried to our rooms to rest our feet! The hours on the concrete and stone had taken their toll and we were glad to take it easy for a while.

The next morning we were up early and on our way. It was still raining heavily, but it appeared to be gradually clearing. We were leaving the wine and grain country and slowly but surely climbing toward the mountains. A detour in Pamplona complicated our getting through the city, and military road blocks and check points told us that someone was wanted for a crime or

My largest workshop project. Ten smiths, ten days to make the arch, twenty-nine feet across, made in three sections and bolted together. Letters are brass. 1980.

that there was a threat of Basque separatist terrorism. Our drive ended descending a river gorge for the last fifteen miles. We were hemmed in by steep walls on both sides, with the highway and the river close together. It was like driving down a narrow hallway. We had no problem at the border as our red tourist plates must have helped. We got to the hotel early after winding our way down some steep and narrow streets. Su and Stephen were glad to be back in France as the terrorist business had gotten on their nerves. And myself, after taking 400 slides and tramping all over, my feet and my camera needed a rest. Hendaye surrounded by a high fog reminded me of Carmel, California.

June 17th. The fog was still heavy in the morning, but as I looked out the window, I gazed down on a beautiful back yard garden, all terraced neatly, with all kinds of vegetables growing. Roses lined the fences and it was obviously an extremely useful and well kept home garden. We were in a very quiet part of Hendaye, with only one through street and we greatly enjoyed the peace and quiet of the day. We had an excellent dinner that night after an eventful day and had a good night's rest.

At our comfortable hotel in Bordeaux we took time to reminisce, do laundry, and write cards. We recalled that in one of the cities we had visited the specialty had been roast suckling pig. We had been the first in the dining room and somehow had caught the fancy of the owner. He patted me on the shoulder with the "Buenas Noches" greeting as he sat down for his dinner before the ten o'clock rush. At the end of our meal, our waiter brought us a huge brandy snifter, which took two hands to hold, and after washing it out with hot water, poured us an after dinner drink. SOME after dinner drink! At Burgos, it was roast leg of lamb, done on charcoal. At Hendaye, we had been served with a champagne aperitif, to celebrate the end of our brief tour of Spain. The champagne was flavored with currant juice, and for petit dejeuner, a currant compote.

I found it difficult to change so quickly from one country to another. Here in the states, the customs and the language are pretty much the same, but over there, things were quite different. My Spanish, acquired during the Monterey-Carmel days, was just becoming familiar, when we crossed the border into France and I had to begin all over again. Stephen is fluent in both, Su almost as good, so they took care of me in the language department. How will my refreshed German do in Switzerland?

The fog was heavy leaving Hendaye. The country side was sand dunes and pines, it reminded me of summer on the California coast. The toll road had a speed limit of 80 mph, so we zipped merrily on to Bordeaux. In the afternoon, I walked for an hour up and back on one street. I found several examples of ironwork in railings and grilles on the way and eventually came to the shop of a metalworker. Alas, I found the same scene that I had so many times before, the chop saw, Hossfeld bender, and the arc welder. There was an unused forge and anvil in one corner of the shop sadly dusty and deserted. The scene was depressing to me.

In the evening we attended a classic French dinner at the Quintard's. Michel Quintard is the head of the department where Stephen had been doing his research. The dinner was underway when we arrived. A table was set on the lawn covered with hors d'oeuvres, paté, melon, tomatoes, ham rolls, and a bottle of Sauterne. Later when we sat down to dinner inside we had another wine, a monkfish loaf, cold, and another wine, followed by roast quail, a different wine. This was all followed by potatoes and mushrooms and yet another wine. A total of five wines for each course of the meal—after all we were definitely in the best wine country in the world. Dessert was a chocolate cake floating on an island in champagne followed by coffee with cognac. Dinner ended at 11:15! Pete Berg, one of Stephen's classmates in college, whom I had not seen for forty years, brought an additional bottle from his own winery in California, that made SIX!

June 18th. This was a rest up day and we needed it. I was up late and on my own until dinner. I took a late breakfast and then took a walk around town. I found the old Hotel de Ville, which was now a number of municipal offices, but it had a lovely park surrounding it. Su came by later to take me on the bus downtown. We got off the bus and took a long walk through the old city of Bordeaux. We had a small comedy of errors on meeting for dinner. Su, the Bergs, and I waited at one bar until past eight o'clock, while Stephen waited (for us) at another bar just two blocks away. We finally got together at 9:30 and had a good laugh and a rather late dinner. We didn't get to bed until past midnight. Two nights in a row of after midnight were more late hours than I was used to.

June 19th. We were on the road to Carcassonne by 1:00 after some much needed rest. Su and Stephen were prepared for another six weeks of travel after I was to return home, so the car was loaded. The French toll roads were certainly beautiful to travel on. Double guard rails in the center, single guard rails on the outside and there were planted shrubbery and trees on either side of the roadway. It was like driving through a green house all groomed and beautiful with the mountains on either side. There were frequent rest stops with gas stations and various amenities, one even had a hotel.

We arrived in Carcassonne on schedule. There were no cars allowed on the streets in this historic city between 7:00 a.m. and 6:00 p.m.—a great idea—so we followed a hotel van to a parking lot, left the car, and loaded the baggage into a van named "Rascal," and were driven to the Hotel Donjon. It made for a very pleasant city with no cars during the day and we planned an exploring trip for the next day. We had an excellent dinner at 7:00 p.m. and off to an early bed.

June 20th. This turned out to be a most interesting day. The petit dejeuner was the most lavish in a long time. We had fruit, cereal, cold cuts, eggs, etc. We decided to visit the famous castle, which was a huge landmark there, and took the French language tour which was much smaller than the English tour. We discovered a beautiful classic iron plated door to one of the main chambers that was worth the trip. The tour included a long walk on the inner walls, and then we toured the area between the inner and outer walls. The tour between the walls highlighted the different kinds of stone work representing the different construction periods of the huge castle and its long history. Our history in America is only the twinkling of the eye to the long history of Europe.

After the Castle we toured the ancient city. The old downtown is so small that one cannot get lost, just keep making right turns and one soon comes back to the starting point. I wish to emphasize the pleasure of being in a city with NO cars. The tranquility is such a contrast to the frantic traffic in the other cities we had visited.

June 21st. Our last evening in Carcassonne, and our last evening with Stephen, who is returning to Bordeaux, and to rejoin us in Zermatt. Su chose a restaurant where travelling musicians performed. The trio was from Ireland. What a mixture, in a centuries old French city, to have Irish singers in their own tongue, as well as French and English, entertaining us while we had our dinner. The music was very low key. They had no amplifiers and the songs were humorous and very entertaining.

June 22nd. Stephen got us out of the city easily and to the train station. We saw a few interesting bits of iron, but the most interesting sight was watching a small boat go through the locks on the river. The gates were hand operated, as they had been for hundreds of years, and it was fascinating to see the boat slowly drop out of sight in the deep lock.

French trains are all electric, fast, and on time. To Marseille first, and then a train change to go to Aix-en-Provence. It is hardly the small town that I had visited sixty-five years ago. Now it is bustling with young students and many young travellers on the go. A week long music festival was in full swing and each square had an extra loud jazz or rock band. It took us an hour to find a quiet little hole in the wall restaurant for lunch. Fortunately our hotel is on a side street and our rooms open on to a lovely inner court. Peace and quiet prevail here far from the music festival.

June 22nd. The old inner town of Aix-en-Provence has been surrounded by urban sprawl. The ironwork of the old town had its own style as many of the towns we had visited. There were huge ground floor grilles, 7/8" or 1" square set on the diamond. The lower part projected out a foot or two, and then sloped back to the wall at the top. There were lots of wavy water bars, alternating with straight ones. The classic elaborate balconies with scrolls and repousse leaves are everywhere. The sky was grey and I was afraid my photographs were not going to be very good. A light rain is falling.

There is one blacksmith listed in the phone book, but after a call to his shop, I found that he was out of town. I would try to get a sampling of the ironwork in town from the postcards or tourist literature.

My thoughts go back to Spain, perhaps these words apply to not only to Spain and France, but to many other countries. "The art of forging is dying in the place of its birth." Perhaps the final nail in the coffin was not driven today, but my travels had revealed that the lid was certainly ready to be nailed shut.

For some time, in several cities, I had been puzzled by an odd detail in the horizontal bars of the many grilles that I had seen. In the process of splitting a 3/4" square bar to pass another bar through it, the result was that the outside corners of the split were rounded, as in bending a piece in a vise. The corners in all these horizontal bars were as sharp as an upset square corner. Also, in a number of the cases, a small slot remained at one corner of the square hole. Was this a casting? The work did not look cast. Was a longer split made and the corner upset. I was baffled. There were no hammer marks anywhere. Then I saw one, unpainted and close at hand in full sunlight. The bars were flame cut. NO mistaking the signs. The tiny beading that the preheat flame leaves, the very sharp underside of the cut, and the very fine vertical lines that a machine cut leaves. This explained why there were no hammer marks. The small slot at one corner was to avoid the difficulty of piercing a hole in the 3/4" plate.

In some cases a plug had been inserted in the slot and arc welded in place. In this manner, one of the oldest techniques, piercing a bar to pass another one through, had fallen victim to the cutting torch and the arc welder. I will search no further for blacksmiths.

June 23rd. This proved to be a very long day of travel. We entrained at Aix-en-Provence at 8:28, on the first of four trains today. Up we climbed through the foothills to Veynes, where we took a strange switch to get on the RED cars We had been on the YELLOW cars, but they were going to another place. A spectacular trip over the Gorges and into the tunnels to Grenoble. We had a two hour wait between trains so we took time for lunch and a short walk. The short trip to Geneva was through very lovely country and then we switched to another train for our short trip to Bern. I'll never forget the ride along the lake from Geneva. We slowly climbed past hillsides sloping steeply to the lake. The hillsides were carefully tended and terraced for grapevines. What loving care had gone into the use of the land.

Arrived at Bern at 8:00 p.m. Su got all my train reservations for the final trip to Rotterdam. Train tickets and seat reservations are separate items, tickets were purchased in the U.S., but seat reservations could only be made within a certain time period. I am all set for the final leg of my wonderful journey.

After three weeks of relying on Su and Stephen in France and Spain, I am pleased that my German has come back so quickly, thanks to a refresher course at CRMS. I could order meals, buy stamps, get directions to a camera shop and order a new battery, all with no problems. Su and I bought a fine pocket watch for Stephen's birthday. I found some superb ironwork in Bern. The many grilles on the Post Office were eight feet wide and twenty feet high. There were gates, grilles, and balcony railings everywhere.

KEYS! There are many different ways to handle room keys. They are all attached to a very heavy metal tag (so that you don't become too attached to them) and too heavy to carry around. You can leave them at the desk at some hotels, and learn to ask for them in three languages. At one hotel they key was simply hung on an open key board, but that gave me some qualms. In Bern, there was a simple solution. The keyboard is double, a small key hangs above the place for the room key. The occupant turns the top key and the room key is locked into the key board and you then put the little key into your pocket. When you return, you insert the small key again and the larger room key is released and you can get into your room. It is a great system, simple, convenient, and very handy. The Swiss are a very clever and inventive people.

June 25th. Yesterday was a full day of walking. Su and I did at least four miles on our shopping jaunt and I added another mile after lunch. First a watch for Stephen, easily found at the first shop. Pocket watches are scarce here because people just don't use them much any more, but the one we found was an unusually fine one. It had a quartz movement, a second hand, and a date display. It was perfect and just what we wanted.

We continued our walk across one of the many high bridges in the very old town, then down at least 200 steps to a lovely walk along the river. We stopped at a restaurant were Stephen and Su had had lunch seven years ago. It was worth the walk. We had a fine lunch as we watched the water spilling over a flood control gate.

After lunch we headed back up the steps toward the Cathedral to hopefully view some very fine ironwork. There are four possible ways to find ironwork in Bern. Many of the streets have arcades over the sidewalks, so we walked in both directions on both sides of the streets. Here and there we found what we were looking for.

Later we stopped in front of the Albert Einstein house, which was unfortunately closed. We had an excellent dinner at our hotel, The Goldener Schlussel, and heard a good story about Heaven and Hell, pertinent to the mixture of nationalities in Switzerland.

Heaven		**Hell**
English	Police	German
French	Cooks	English
German	Mechanics	French
Swiss	Bankers	Italian
Italian	Lovers	Swiss

June 25th. Off to Murren this morning. It took forty-five minutes at the train station to get Swiss Rail Passes validated and final seat reservations. It took two hours to Interlaken, then a short ride on a train with cog wheels, then a final ascent of thirty-five degrees up two thousand feet on another cog rail. We had a few glimpses of the Alps on the way, but the clouds were heavy. As we settled down for lunch, the clouds rolled in solid and the visibility became zero.

June 26th. In Murren, we took a four mile walk to the trolley station at the head of the inclined cog railroad. The visibility was still zero when we started, but it slowly lifted. There were steep hillsides which were used for fields with cows grazing, and we counted thirty varieties of wild flowers. The timber on the hillsides was very thick and we met only two cars on the way. We caught the trolley back to Murren. That afternoon we took a short nap and then searched the stores for postcards and took a walk to find our trail for tomorrow.

We had a fine dinner that night overlooking a deep canyon, and while we were eating, the clouds really rolled in again. Just before 9:00 p.m. the clouds lifted and we could see the famous Eiger Wand, rising 11,000 feet above the valley. It was spell binding, as the clouds slowly cleared. I sat for half an hour watching the ever changing scene. It was spectacular.

June 27th. Hooray! A crystal clear morning had dawned with a 180 degree view of the magnificent crest. We decided to walk to Grimmelwald, a tiny village 300 feet below, we thought. We forgot that the contours on the map were in meters instead of feet, so it was really 1,000 feet down. There was a tremendous network of trails and walking paths, linked by an equally planned series of gondolas. As the morning progressed, clouds formed on all the peaks, so that by the time we got to the

Schilthorn, we were shrouded in fog. We saw several paragliders taking off from the halfway point on the gondola ride. They had 7,000 feet of clear free air under them to the bottom of the valley floor. What nerves, or lack of them!

In my time, I have seen many mountain ranges. The Green Mountains of Vermont, the California Coastal Ranges, and the Sierra Nevada, the Rockies of Colorado, the Canadian Rockies, the Appalachians and the top of Georgia, the Pali in Hawaii, a bit of Alaska, and the southern New Zealand Alps. Each range has its own character and quality, but here, perched on a ledge on the hotel balcony halfway up the heights, I can meet the mountains face to face. The sheer wall of the Jungfrau cannot be more than a quarter of a mile away. No foothills, no long approach, there is nothing in my memory of fourscore years in the mountains that equals this experience. It is certainly awe inspiring to say the least.

June 28th. It was sad leaving Murren, such a lovely place, and no cars. We started another long day. We took the trolley to the cog rail and then on to Zurich. We changed trains there and on to St. Moritz. Su said the hotel was only a few blocks from the train station, but as we pulled into St. Moritz, it was right next door. We walked down to the tunnel and right into the hotel door. The hotel was clean and comfortable, but the wind blew cold from the lake. Tomorrow we would ride the Glacier Express.

June 29th. The Grande Finale to my European trip began the night before and in the early morning hours. My hotel room overlooked the train station and train yard. I have thought of my brother Rogers many times on this trip, and this morning I was thinking of him and how he would have loved to watch the train being made up and departing.

My first discovery about the train was that it was a narrow gauge, not quite as narrow as in the U.S., one meter wide instead of three feet, but nevertheless narrower than the standard gauge lines. The width was necessary to negotiate the tight turns in the mountains. It was also a cog railroad. About twenty times a day, the train would slow to a crawl, and we would hear the cogs engage and then the train would continue at a normal speed. Some of the grades were as much as eight percent, twice the maximum for normal tracks. The cogs were for climbing up the steep inclines and for extra braking on the downgrades.

Swiss trains and Swiss tracks are marvels of engineering. The trains cover a network of the entire country and this is certainly NOT normal terrain! You can go anywhere in Switzerland on the train. They are all electric and always *on time.* They have to be to allow passage at the double track sections, one of which was in the middle of the nine mile long Furka Tunnel. Think what a mess that could be if one of the trains was a few minutes late or early!

The same care the Swiss have for their trains, they have for their land. Every square foot is tenderly cared for. Dense forests, many planted by hand, cover the country side. Hay fields flourish on the almost impossible slopes. Grape vineyards thrive on the terraces, and grain and potatoes grow in the valleys. There are home gardens everywhere.

In Murren, there are vegetable gardens

on thirty-five degree slopes and here in Zermatt we are surrounded by hotels on all four sides and each has a large garden next to it.

Haying season was in full swing as we left St. Moritz. On the very steep slopes it was hand cut with scythes. On the gentler slopes they used mowing machines that were like little garden tractors. Most of the crops on the steep slopes were cultivated vertically instead of horizontally. Once we saw a man tethered by a rope from above swinging a scythe in a long arc across the field. We assumed that he used this method so he could concentrate on his cutting rather than trying to stand upright.

I cannot help but compare this train to the Cumbres and Toltec narrow gauge where I placed the Frimbo Memorial plaque in memory of my railroad riding brother. The land along the Cumbres Toltec route is almost uninhabited whereas the route we were taking today was lined with quaint cottages mixed with modern homes.

As I ride along so easily and effortlessly I am thinking of the men who built this line so long ago. With sheer physical power and dynamite they chopped and chiseled this route through impossible terrain and dangers and created one of the wonders of the world. Clinging to canyon walls hundreds of feet high above a rushing stream, coming out of a tunnel and crossing a high bridge, this trip makes the wildest roller coaster ride seem tame by comparison. There were several loop tunnels that bore into solid rock and came out directly above or below the first entrance into the mountain. The nine mile tunnel under Furka Pass carries passenger cars in the winter, as the pass is closed for eight months of the year!

Once in a while, I had to shut my eyes or turn away from the window as we seemed to be gliding through the air and I didn't like the feeling. I knew the rails were still under the train, but it didn't look very reassuring from the window.

On Cumbres Pass, the towering peaks and the glaciers form an equally hair raising experience. I believe it is the most spectacular of all my train trips and a fitting place for the memorial to my brother. He called The Glacier Express the "slowest express train in the world." Rogers loved to travel on the Swiss trains also. I am happy to have followed in his footsteps on The Glacier Express.

The Glacier Express had all the flavor of the old world luxury train travel. The dining car was luxurious and unhurried. We started lunch with a champagne toast to Rogers and his constant companion E.M. Frimbo, ending with another toast with brandy, as close as we could come to his favorite, Brandy Alexanders. It was truly a wonderful day with many memories of days gone by with friends and family.

Stephen met us, having driven within seven miles of Zermatt, there he had to park his car and ride the train. Taxis are the only cars allowed into the area, and then only up to a certain point. Beyond that, small electric carts and wheeled baggage carts are used. The Alphubel Hotel was very pleasant. The rooms were small, but comfortable. We were on pension there with only breakfasts and dinners, but that was enough. The dinners were excellent and Su and Stephen could converse with our waitress in French. I, with my revitalized German also worked in well. The waitress was surprised that we were Americans. She said, "Americans never stay in two star hotels, they only stay in four star hotels." Su told us that was because one can seldom find less than four star hotels with a U.S. travel agent. The two star hotels in Switzerland are excellent.

I found a metalworker in Zermatt, but the story was the same. I could converse with him easily and he told me that they no longer forged things themselves, but bought all their parts, scrolls, balusters, and split baskets prefabricated in Italy. They then arc welded them together. I found the neglected and dust covered forge and sadly took my leave. It was full of cigarette butts!

We left the hotel at 5:30 a.m. to catch the first train. Su and Stephen got off at the parking area and I went on to Brig. There I took the first of four trains to Rotterdam. As I wrote this, I was traveling along the Rhine, which I had followed from its source to the sea. A big broad river valley with a lot of farming and many vineyards on both sides of the river. I had to change trains again, which made five. From Brig, to Basel, to Köln and finally to Rotterdam, where I would have a few days rest before my long flight home. My story is coming to an end, the land is FLAT, and so am I!

July 2nd. Since the Waals-Wilton home is very small, perhaps twenty-five feet wide with very limited space for a family of four, it is their custom to put up guests at a hotel. So, after a brief visit to their home, Piet took me to the hotel they had selected. The modest hotel was very pleasant. It faced a small boat harbor and was in a section of town where there were many fine old buildings. As I had been on the trains for fourteen and a half hours, I was ready to sleep in.

I was in bed by 11:30 p.m., but what Piet (and I) didn't know was that this hotel was in the midst of major remodeling in preparation for the imminent holiday season. Hammering began at 7:30, the jackhammers right above me started at 8:00. So much for sleeping in! I asked the clerk if this was going to be going on much longer, and he said that it would be at least for the rest of the week.

A change of hotels was definitely in order. Four star hotels near the airport or the train station were ruled out as we were at the height of the tourist season and they would have been booked full. So Liesbeth and I went searching, after first picking up my bags. The first hotel we checked on had only a hall bathroom, no elevator, and a dangerous set of steep steps. The next one was full, but the manager gave us a card for another one only a few blocks away.

This one was adequate. It was one large room with a bath, twelve foot ceilings and beautifully decorated plaster mouldings with baroque heads and scrolls everywhere. The high windows had stained glass sections across the tops. Although there was a fairly well-travelled street in front, there was very little traffic noise. Anyway, it was the beginning of the weekend and most of the city people were leaving for the country. Traffic would certainly be light on the weekend.

Monday morning I was to leave at 6:00 a.m. for the beginning of the long flight home. I took a walk around the area on Saturday morning. The street in front of the hotel was about 175 feet wide, this was a sharp contrast to most of the narrow streets in the old cities we had visited. It sported a fifteen foot sidewalk, with a secondary one-way access street, allowing for traffic and parking. There was a generous planting strip down the middle with huge trees that shaded everywhere. There were two lanes of traffic both ways in the center and the pattern of the secondary road and the sidewalk was repeated on the other side.

July 3rd. Liesbeth picked me up at 10:00 a.m. to go on a short visit to the city of Gouda. Gouda is famous for its cheeses. It was easy to see the boundary between the city and the country. You are thrust instantly into lush green pastures and green and white houses. Gouda proved to be a really charming city. It was very old and the main square was dominated by a 16th Century Gothic Cathedral. After a leisurely walk, we enjoyed a light lunch and continued our tour. We saw some nice old ironwork in various places and then returned to Rotterdam for a dinner at their favorite Swiss restaurant. Before we were finished we were treated to a pouring rain with accompanying thunder and lightning which fortunately cleared the air during the night.

July 4th. For the third consecutive year, I find myself out of my native land on this day of the celebration of its birth. I wonder where I will be next year on this date? I am missing the fireworks from the top of Mt. Sopris, and sadly remembering the days when Wagner Park would hold all the residents of Aspen who would come to see the fireworks on Aspen Mountain. Piet took me to The Hague. Remarkable ironwork was visible everywhere. We then went to the Peace Palace, the Shell Oil Building, and finally to the Royal Palace. It was a warm pleasant day, partly sunny, so I will leave on a pleasant note weatherwise. If there is one thing common in all the cities that I have visited, except where the automobile is banned, is that the automobile has almost destroyed the natural historic beauty of the pre-automobile towns. Traffic, air quality, and noise have made them, at least for me, unliveable. Coming down from the Alps at 6,000 feet to below sea level, I found the air for the most part, heavy, oppressive, and at times downright foul.

Back to June 30th. I missed writing about the trip from Zermatt to the Gornergrat. It was a very bright clear day when we boarded the train. We walked a couple of blocks to the station and climbed aboard a car with a sloping floor. I guessed a thirty-five degree slope. We made several stops along the way to pick up and deposit passengers, as many of the Swiss travel on the trains for their local transportation—they are much safer than the roads! We enjoyed magnificent views of the Matterhorn on our climb and the higher we got the more breathtaking they became. There were many glaciers that had been there for centuries and a full circle of towering peaks surrounded the dominating Matterhorn. Across a wide canyon, there was a summer ski slope and we could see the tiny dots of skiers coursing their way down the hill sides. We had a snack on the top, sharing our food with some hungry crows who had stationed themselves strategically for the tourists. Then we started down enjoying a different view of these fabulous mountains, but equally as breathtaking.

July 5th. Twenty-three hours from one hotel in Rotterdam to the next one in Grand Junction, but I felt it better to make it all in one jump rather than spending another night on the road. My journey had ended, but the memories and the pictures will last for the rest of my life. All in all an incredible journey back into history and into the many cultures of Europe.

9

KEEPING THE FIRES BURNING

When the Artists and Blacksmith's Association of North America was formed at Lumpkin, Georgia, in 1973 from the encouragement of Alex Bealer and others, I was continuing my profession of many years at the Mountain Forge in Aspen, Colorado.

In 1976, Portia and I received an invitation to attend the ABANA National Conference in Carbondale, Illinois and I was also invited to demonstrate. I had heard of the formation of a national association through a friend, but up to that time I had not attended any of their conferences or workshops. I had done some demonstrating to school and art groups when I had lived in Carmel, California, but nothing on the scale of this.

Now for the first time I had before me an audience that had some experience in working with iron and knew some of the rudimentary techniques associated with the craft and art of blacksmithing.

To say I was a hit is probably an understatement. I had the early experience with men like Samuel Yellin in Philadelphia and with Julius Schramm in Germany plus a lifetime of experience working at the craft. I realized that I was probably the only living link with the early blacksmithing in America and the present day. My mission for the next twenty years or so became suddenly clear, to keep the fires burning and keep the craft and art alive and growing. I think I have accomplished this goal.

During the conference, Portia and I let it be known that we would be available wherever and whenever a conference or workshop was to be held. The response was instantaneous and has continued to this day.

The following is a list of most of the Demonstrations and Workshops that I have participated in since 1945.

1945–60	School and Art Groups, Monterey County Fair
1964–68	Colorado Rocky Mountain School, Carbondale, CO
1975	Sterling Arts Council, Sterling, CO
1976	ABANA National Convention, Carbondale, IL
1977	Aspen Historical Society, Ashcroft, CO
	Middleton Gardens, SC
	Tullie Smith House Blacksmith Guild, Atlanta, GA
	Dr. Calvin Smith (local group), Greenville, SC
	John C. Campbell Folk School, Brasstown, NC

Robert Hale (local group), Kingston, TN
Emmert Studebaker, Four State Roundup, Tipp City, OH
Goat Hill Forge, Nashville, IN
Southern Illinois University, Carbondale, IL
Southern Illinois University, Edwardsville, IL

1978 California Blacksmith Association, Sacramento State University, Sacramento, CA
Southern Illinois University, Edwardsville, IL
John C. Campbell Folk School, Brasstown, NC
ABANA National Convention, Purchase, NY. Elected ABANA President,
Southern Illinois University, Carbondale, IL
Emmert Studebaker, Ohio Blacksmith Association, Tipp City, OH
Brotman Forge, Lebanon, NH New England Blacksmith Association
California Blacksmith Association, Teaching Symposium

Invitations

1979 International Medieval World Conference, University of Wisconsin, LaCrosse, WI
California Blacksmith Association Annual Conference
Workshop, Harpers Ferry, WV
John C. Campbell Folk School, Brasstown, NC
Haystack School, ME
Stuhr Museum, Grand Island, NE
Peters Valley School, Layton, NJ
Southeast Blacksmith Association, Savannah, GA
Appalachian Blacksmiths Association, Cedar Lake, WV
Appalachian Chapter ABANA, Chattanooga, TN
Ashokan Campus, State University of New York, Kingston, NY
Philadelphia College of Arts, Philadelphia, PA
Appalachian Center for Crafts, Cookeville, TN

Master Class Architectural Projects Conducted by Francis Whitaker

1979 University of Southern Illinois, Carbondale, Illinois, (Gates and railing Faner Gallery)

1981 Fort Worden State Park, Port Townsend, Washington (Railings for Officers Quarters Restoration and Entrance Gates for Fort Worden State Park)

1983 Samuel Yellin Museum, Philadelphia, Pennsylvania (Memorial Gate for the Museum)

1983 John C. Campbell Folk School, Brasstown, North Carolina (Entrance Railing, Porch Stair Railing)

1983 Constitution Hall Park, Huntsville, Alabama (Railings for Restoration of Constitution Hall)

1983 Stone Mill Winery, Cedarburg, Wisconsin (Upper Midwest Blacksmiths Entrance Archway)

1984 California Blacksmith's Association, San Martin, California (Garden Gate)

1984 Heritage Square, Phoenix Arizona (Arizona Artist Blacksmiths Archway for Heritage Square)

1984 By Hammer and Hand, Rowe, New Mexico (Driveway Gate)

1984 Artist Blacksmith's Association of North America Conference, DePere, Wisconsin (Gate for National Ornamental Iron Museum, Memphis, Tennessee) These Master Classes involve six to twelve advanced smiths, and take a major architectural project from design, layout, forging, assembly, and sometimes installation

1985 City of Woodland Park, Colorado (Twenty-nine Foot Archway for a City Park) Completed September, 1985. Smiths from four states participated in the master class workshop.

1986 Second Gate Workshop, Samuel Yellin's, Philadelphia, Pennsylvania

1990-93 Three chandelier making workshops at the John C. Campbell Folk School, Brasstown, North Carolina

Workshop Schedules

1979 Stuhr Museum, Grand Island, NE
Peters Valley School, Layton, NJ
Southeast Blacksmith's Association, Savannah, GA
Appalachian Blacksmith's Association, Cedar Lakes, WV
Appalachian Chapter ABANA, Chattanooga, TN
Ashokan Campus, State University of New York, Kingston, NY
Philadelphia College of Arts, Philadelphia, PA
Appalachian Center for Crafts, Cookeville, TN

1980 Sitka Community College, Sitka, AK
Unicorn Forge, Barneveld, WI
Cedar Creek Forge, Cedarburg, WI
Peaceable Kingdom School of Blacksmithing, Washington, TX
Campbell Folk School, Brasstown, NC
Hale Blacksmith Shop, Kingston, TN

1981 Iron Anvil Workshop, Castle Rock, CO
Northwest Blacksmith's Association, Port Townsend, WA
Sitka Community College, Sitka, AK
Haystack School of Crafts, Deer Isle, ME
State University of New York, Kingston, NY
Studebaker Workshop, Tipp City, OH
Cedar Creek Forge, Cedarburg, WI
Campbell Folk School, Brasstown, NC
Greenville Museum of Art, Greenville, SC

1982 Northwest Blacksmith's Association, Port Townsend, WA
Southwest Blacksmith's Association, Monroe, NM
Cedar Creek Forge, Cedarburg, WI
Pehoski School of Blacksmithing, Salado, TX
Campbell Folk School, Brasstown, NC
Arizona Artist-Blacksmith's Association, Phoenix, AZ

1983 Philadelphia College of Arts/Samuel Yellin Foundation, Philadelphia, PA
Gateway Farriers, St. Louis, MO
ABANA Southeast Conference, Madison, GA
Constitution Hall Workshop, Huntsville, AL
Campbell Folk School, Brasstown, NC, Spring/Fall
Southwest Artist Blacksmiths, Rowe, NM
Cedar Creek Forge, Cedarburg, WI

1984 California Blacksmith's Association, San Martin, CA
J.R. Harkins Workshop, Carmel, CA
Northwest Blacksmith's Association, Rochester, WA
Missouri Blacksmith's Association, Bethel, MO
Indiana Blacksmith's Association, Noblesville, IN
John C. Folk School, Brasstown, NC
Appalachian Center for Crafts, Smithville, TN
Southwest Artist-Blacksmith's Association, Rowe, NM
ABANA Conference, DePere, Cedarburg, WI
Ashokan Field Campus, Kingston, NY
John C. Campbell Folk School, Brasstown, NC
Arizona Blacksmith's Association, Phoenix, AZ

1985 John C. Campbell Folk School, Brasstown, NC
Rocky Mountain Smiths, Carbondale, CO
Rocky Mountains Smiths, Aspen, CO
Woodland Park Arch, Woodland Park, CO, Master Class

1986 John C. Campbell Folk School, Brasstown, NC
Samuel Yellin Metalworker, Philadelphia, PA, Master Class Gate
Southern Ohio Blacksmith's Association, Tipp City, OH, Master Class Gate
Blacksmith's Association of Missouri, Owensville, MO, Master Class Grilles
By Hammer and Hand, ROWE, NM, Master Class Gates
Artist Blacksmith's Association of North America, National Conference, Flagstaff, AZ, Demonstrations

John C. Campbell Folk School, Brasstown, NC

1988-93 Resident Artist Blacksmith at Colorado Rocky Mountain School, Carbondale, CO

1992-94 Two Master-Apprentice Grants from the Colorado Council on the Arts

1993 Final Class at John C. Campbell Folk School, Brasstown, NC

First Demonstration at Allison's Wells School, Canton, MS

Final Demonstration at Nicholas Brumder's, Georgetown, TX

Honorary Life Member

Upper Midwest Blacksmith's Association
Appalachian Area Chapter ABANA
Arizona Artist Blacksmith's Association
California Blacksmith's Association
Rocky Mountain Smiths
Mississippi Forge Council
Tullie Smith House Guild, Atlanta Georgia
New England Blacksmith's Association
Alabama Forge Council

Past President

Artist Blacksmith's Association of North America
Southwest Artist Blacksmith's Association

In 1985 I took a sabbatical year from workshops and wrote a textbook for advanced blacksmiths, *The Blacksmith's Cookbook, Recipes in Iron*, published by Jim Fleming Publications of Vail, Colorado. All proceeds from the book have gone into a non-profit, tax-exempt, educational foundation, The Francis Whitaker Blacksmith's Educational Foundation. The primary purpose of this foundation is to conserve the principal and use the annual interest to fund scholarships for advanced smiths, loans for funding, conferences, and to document significant ironwork.

I dedicated the book to Samuel Yellin. It was at his forge in Philadelphia that my career began, in 1922 at age 15. A year at Yellin's, two more at Julius Schramm's in Berlin, Germany, three years of intensive training under two great masters, gave me the background knowledge to go on and improve my techniques, and develop my own style of ironwork.

It has been my intent "to keep the fires lit" in the blacksmithing profession since my first ABANA conference in Carbondale, Illinois, and I feel secure in the fact that I have done just that. My personal message has reached thousands and the message is still going out.

I am especially proud of my contribution to the John C. Campbell blacksmithing facility. When I first started going there for demonstrations in 1977, the shop was nothing but an open shed, now there is a fine well equipped shop with forges and work stations for twelve students. Quite a change from the old days. Also it has been named the Francis Whitaker Blacksmith Shop in my honor.

When my wife became ill with terminal cancer, my traveling partner and helpmate was suddenly not able to help any longer. She held on for a year and a half, but we all knew the end was not far off. When she passed away, I decided to greatly reduce my speaking engagements and demonstrations and confine my activities to the Rocky Mountain School and my home shop. The people that I had gone out to for so many years could now come to me. We have had several very successful blacksmith conferences at my home shop in Carbondale and I hope to continue to do so. A number of smiths have come to me for advanced training with their own architectural projects.

10

AWARDS AND ACCOLADES

I never really sought recognition in the early years. I was too busy with the daily struggle for survival for myself and my family. However, when the depression was over and the pressures of just making a living were lessened, I began to look outside the walls of my own shop into the business of the community in which I lived.

Carmel in those days was a thriving, building community and in some ways the building was affecting what I thought was a wonderful place to live and raise a family.

In 1952 I entered politics for the first time. I ran for the Carmel City Council, and to my surprise and the surprise of many others—I was elected. I served on the council from 1952 until 1963, being reelected several times. I was also on and elected to the presidency of the Carmel Craft Guild in the early 1960s.

I was at that time, and continue to be, interested in the preservation of our natural environment and from that interest I joined the Point Lobos League. I eventually became president of that organization as well. We were successful in establishing the first marine reserve in the United States surrounding Point Lobos—a beautiful point along the Pacific coast.

Upon moving to Aspen I continued my interest in city government and served on the Aspen Planning and Zoning Commission and was elected chairman of that organization in the mid-1960s.

I also served on the Aspen City Council from the mid-1960s to the early 1970s. At that time I was also a member of the Aspen Charter Commission, and the Aspen Board of Adjustments.

In 1976, with the help of my wife, Portia, I received an Honorary Doctorate of Humane Letters from the University of Colorado. Also I was named an Honorary Life Director of the Pitkin County Parks Association.

My honors in my chosen professional field came slowly, but lately there has almost been an avalanche of them.

In 1978 I was elected President of the Artist-Blacksmith Association of North America. Over the years between 1945 and the present, I have been the demonstrator of over one hundred and fifty blacksmith workshops from Florida to Alaska and most points in between. In 1990 I was named a Fellow of the College of America Craft Council. This year (1993) the Blacksmith School at the John C. Campbell Folk School was named in my honor, The Francis Whitaker Blacksmith Shop, in recognition of my two decades as a demonstrator and instructor there.

May 1, 1993 was proclaimed Francis Whitaker Day in Canton, Mississippi in recognition of my contribution to the preservation of blacksmithing in that state. My home town of Aspen had given me a similar honor in 1988.

This summer I placed an exhibit in the National Ornamental Museum, in Memphis, Tennessee, with an accompanying award for Lifetime Achievement.

I also placed an exhibit in the Faner Gallery on the campus of the University of Southern Illinois in Carbondale, where I was named an Honorary Fellow for contributions to the art of blacksmithing. A small exhibit will be displayed in the fall of 1993 at the University Gallery, Radford University, Radford, Virginia.

I think one of my most lasting contributions may be my authorship of *The Blacksmith's Cookbook*. I am sure this book will be in demand for many years to come and has a great deal to offer the beginning or the advanced blacksmith.

With my move from Aspen I founded the Francis Whitaker Blacksmith School in Carbondale, at the Colorado Rocky Mountain School. This complete facility had served as an educational tool for spreading the information and inspiration to blacksmiths from all over the United States. Yes, some perspiration, too!

With the proceeds from my book and from other sources and contributions, I founded the Francis Whitaker Blacksmith's Educational Foundation. Funds from this foundation can now be used to sponsor promising smiths and to fund workshops which will improve the quality of the profession for many years to come.

As you can see, I have been involved my entire life in the preservation and promotion of my profession, preserving the environment in which I lived, and carrying on those principles of the pursuit of perfection in the activities of my daily life and my professional life.

My greatest reward in life is seeing the profession of the Blacksmith not only survive, but flourish into the twenty-first century.

Accolades and Honors

Three times elected to the Carmel City Council, 1952–1963.

President of The Point Lobos League, helped establish the rules for Point Lobos Reserve, and the first under water reserve in conjunction with the Reserve.

Member of the Monterey Bay Area Planning Commission.

Elected twice to the Aspen City Council.

Served on the Aspen City Council Planning and Zoning Commission, the Aspen Charter Commission, and the Aspen Board of Zoning Adjustments.

Elected President of the Artist Blacksmiths Association of North America, 1978.

Honorary member of numerous blacksmith associations.

The Blacksmith School at the John C. Campbell Folk School, in North Carolina has been named after me, in recognition of seventeen years of teaching and helping improve the shop.

Founder of the Francis Whitaker Blacksmith School at the Colorado Rocky Mountain School in Carbondale, Colorado.

elected to the College of Fellows of the American Craft Council. Just exhibited at the National Ornamental Museum, Memphis, Tennessee in an exhibit for Fellows in Metal.

Received an Honorary Degree in Humane Letters from the University of Colorado in Boulder, Colorado.

Author of *The Blacksmith's Cookbook,* a textbook on blacksmithing techniques.

Founder of The Francis Whitaker Blacksmith's

Educational Foundation, dedicated to the preservation of blacksmithing via scholarships and grants for educational purposes.

Received two grants from the Colorado Council on Arts and Humanities for a master-apprentice program.

Recognized as the "Dean of American Living Blacksmiths."

Nominated for status as a "Living Treasure" under a program of the National Endowment for the Arts.

John Steinbeck used me as a role model in "The Long Valley."

Leon Uris did the same for "Trinity."

Subject of a short documentary on "Tom Brokaw's NBC Nightly News."

1995 recipient of the Colorado Governor's Award for Excellence in the Arts.

The Blacksmith School at the John C. Campbell Folk School, Brasstown, North Carolina, now named after me. 1993.